The Politics of Urban Sustainability Transitions

Cities, the world over, are increasingly recognised to be both a principal source of the environmental and social sustainability challenges facing contemporary society and a critical site for addressing these challenges. Socio-technical systems are at the heart of these challenges as they configure central aspects of urban life: from mobility and energy infrastructures to leisure activities and patterns of mobility. This observation has led to substantial interest in how societies might initiate and actively steer radical transitions in these systems in the pursuit of sustainable urban futures.

This book contributes to emerging debates on the politics of urban transitions by examining the intimate interlinkages between knowledge, power and governance. Drawing upon real-world examples of urban governance, the authors explore the strategies, struggles and controversies involved in configuring knowledge and how knowledge constructions influence governance by rendering some concerns and issues visible and valuable, while obscuring others. The book draws attention to how novel ways of conceptualising, knowing and observing socio-technical systems may be harnessed productively in redefining the power relationships underpinning unsustainable practices. Understanding these dynamics can ultimately inform and enable new approaches to support much-needed urban transitions.

This book provides a compelling examination of urban knowledge politics for the twenty-first century that will be of great value to academics, policymakers and practitioners working in the social sciences, urban studies, geography, urban governance or sustainability transitions.

Jens Stissing Jensen is Assistant Professor, Department of Development and Planning, Aalborg University, Denmark.

Matthew Cashmore is Professor in Environmental Planning at the Norwegian University of Life Sciences, Norway.

Philipp Späth is Senior Scientist/Research Group Coordinator at the Institute of Environmental Social Sciences and Geography, Albert-Ludwigs-University, Freiburg, Germany.

Routledge Research in Sustainable Urbanism

This series offers a forum for original and innovative research that engages with key debates and concepts in the field. Titles within the series range from empirical investigations to theoretical engagements, offering international perspectives and multidisciplinary dialogues across the social sciences.

Co-Producing Knowledge for Sustainable Cities
Joining Forces for Change
Edited by Merritt Polk

Global Garbage
Urban Imaginaries of Waste, Excess, and Abandonment
Edited by Christoph Lindner and Miriam Meissner

The Experimental City
Edited by James Evans, Andrew Karvonen and Rob Raven

Imagining Sustainability
Creative Urban Environmental Governance in Chicago and Melbourne
Julie L. Cidell

Regenerative Urban Design and Ecosystem Biomimicry
Maibritt Pedersen Zari

The Politics of Urban Sustainability Transitions
Knowledge, Power and Governance
Edited by Jens Stissing Jensen, Matthew Cashmore and Philipp Späth

www.routledge.com/Routledge-Research-in-Sustainable-Urbanism/book-series/RRSU

The Politics of Urban Sustainability Transitions

Knowledge, Power and Governance

Edited by Jens Stissing Jensen, Matthew Cashmore and Philipp Späth

Routledge
Taylor & Francis Group
LONDON AND NEW YORK

earthscan
from Routledge

First published 2019
by Routledge
2 Park Square, Milton Park, Abingdon, Oxon OX14 4RN

and by Routledge
52 Vanderbilt Avenue, New York, NY 10017

First issued in paperback 2020

Routledge is an imprint of the Taylor & Francis Group, an informa business

British Library Cataloguing-in-Publication Data
A catalogue record for this book is available from the British Library

Library of Congress Cataloging-in-Publication Data
A catalog record has been requested for this book

ISBN 13: 978-0-367-66487-9 (pbk)
ISBN 13: 978-1-138-47965-4 (hbk)

Typeset in Goudy
by Wearset Ltd, Boldon, Tyne and Wear

Contents

Figures

Tables

Contributors

Wouter Boon is Assistant Professor at the Copernicus Institute of Sustainable Development, Utrecht University in the Netherlands. His research focus is on the role of users and bottom-up processes in technological development and innovation.

Vanesa Castán Broto is a professorial fellow at the Sheffield Urban Institute, University of Sheffield. Her research focuses on climate change politics and urban governance, urban sustainability transitions and insurgent planning practices to address climate change.

Matthew Cashmore is Professor in Environmental Planning at the Norwegian University of Life Sciences, Norway. Matthew is an environmental social scientist and his research addresses the politics of governance arrangements enacted in response to perceived environmental, climate and sustainability crises.

Matthew Cook is Professor of Innovation at the Open University, United Kingdom. His research interests are in socio-technical innovation in cities and associated sectors such as utilities. Much of his research focuses on the governance and management of innovation to achieve more sustainable futures.

Louise Guibrunet is a post-doctoral fellow at the Research Institute on Ecosystems and Sustainability, National Autonomous University of Mexico. Her research focuses on environmental politics, exploring the environmental justice implications of different environmental governance models and sustainability policies.

Evelien de Hoop is a post-doctoral fellow in science and technology studies (STS) at the Copernicus Institute of Sustainable Development at Utrecht University in the Netherlands. Her current research focuses on knowledge politics in the context of urban sustainability experimentation and co-creation.

Ralph Horne is Professor of Geography and Deputy Pro-Vice Chancellor, Research in Design and Social Context at RMIT University, Australia. His research interests span low-carbon transitions in housing and households and the spatial, material and contingent social and policy structures at play in the making and shaping of urban environments.

Shivant Jhagroe is Assistant Professor at the Institute of Public Administration at Leiden University, the Netherlands. His research interests include governance and politics, particularly in the fields of (urban) sustainability transitions and digitalisation/big data.

Ulrik Jørgensen is Professor of Sustainable Transitions at Aalborg University, Denmark, and leads the Center for Design, Innovation and Sustainable Transition at the campus in Copenhagen. His research has been focused on engineering and design competence and, in recent years, science and technology studies, with emphasis on energy systems and transitions theory.

Peter Karnøe is Professor in Innovation and Sustainable Development at Aalborg University, Department of Planning and Development, and is associated with the research Center for Design, Innovation, and Sustainable Transition. He works currently with market and valuation studies and studies sustainable transition as the onto-epistemic work associated with shifting valuations of the political economy of fossil fuels to the making of a low-carbon political economy.

Derk Loorbach is Professor of Socio-economic Transitions and Director of the Dutch Research Institute For Transitions (DRIFT) at Erasmus University Rotterdam, the Netherlands. His action research combines the theory building on governance of sustainability transitions with applied research and consulting for governments, business and civil society to advance and accelerate transformative change.

Rachel Macrorie is a post-doctoral research associate at the Urban Institute, University of Sheffield in the United Kingdom. Her research interests centre around the governance of urban life for environmental sustainability and societal equitability, with a particular focus on resource management, housing construction, automation and robotics.

Simon Marvin is Director of the Urban Institute at the University of Sheffield. His research examines a wide range of issues, including technology, infrastructure, mobility and sustainability, from the perspective of urban planning.

Nanja Christina Nagorny-Koring is Project Manager for European educational projects in the climate sector in the Center for Industry and Sustainability at Provadis School of International Management and Technology in Frankfurt/M, Germany. Previously, she researched questions of the economisation of international climate policy and policy learning in municipal climate governance at the Department of Human Geography of Goethe University Frankfurt/M, Germany.

Stephen Potter is Professor Emeritus of Transport Strategy, undertaking work on the design and innovation processes involved for the diffusion of cleaner transport technologies and more sustainable travel behaviour, including electric vehicles and smarter transport.

Rob Raven is Professor of Sustainability Transitions at Monash Sustainable Development Institute at Monash University in Australia and at Copernicus Institute of Sustainable Development at Utrecht University in the Netherlands. His current research is focused on analysis of transformative change in the urban context with particular attention to the interplay between urban experimentation, institutional change and path-dependent socio-technical regimes.

Adrian Smith is Professor of Technology and Society at SPRU, University of Sussex. His research considers the politics and governance of innovation for sustainability. He is currently developing research into grassroots digital fabrication. He is also involved in a European project studying the knowledge politics involved in smart urbanism.

Philipp Späth is Senior Scientist/Research Group Coordinator at the Institute of Environmental Social Sciences and Geography, Albert-Ludwigs-University, Freiburg, Germany. His main research interest lies in urban and multi-level processes of environmental governance. Currently, he is examining how the promise of an objective, efficient management on the basis of big and real-time data transforms the ways in which urban infrastructures are governed and developed.

Jens Stissing Jensen is Associate Professor, Department of Development and Planning, Aalborg University. His research focuses on how alternative ways of defining, delineating and valuing societal systems may be established, and how such alternative conceptions may contribute to shape sustainable transition pathways.

Alan-Miguel Valdez is a post-doctoral research associate exploring urban-scale innovation experiments at the Open University, UK. Through his career, his research has focused on how experimental spaces facilitate the development of new knowledges about sustainable innovation, and on how such knowledges coalesce and travel.

Preface

The genesis of this book was discussions Jens and Matthew initiated amongst a small group of like-minded individuals that had coalesced around Copenhagen. These discussions took place in a network called 'Urban Innovation and the Governance of Sustainability Transitions', and through our discussions we started to question various dimensions of the linkages between transitions thinking and urban governance. These discussions evolved into the organisation of a conference session in 2014 at the Royal Geographical Society's Annual International Conference, at which point we enrolled Philipp in our ongoing machinations.

We jointly organised a workshop in Copenhagen in November 2016, entitled 'the epistemic politics of urban sustainability transitions'. Our call for papers produced a strong response, particularly from across Europe, and we then had the challenging task of selecting a relatively small number of abstracts for inclusion in the workshop programme. A second, follow-up workshop was held in 2017. Writing can be a lonely, isolating exercise. We are very glad, therefore, that we had a highly engaged group of authors working with us on this book. The workshops were instrumental in generating vibrant discussions amongst the authors and for this we are particularly indebted to the Velux Foundation, who provided funding for the workshops through the research project 'ECOMAC: Ecological macroeconomics and sustainable transition'. Our gratitude extends also to Aalborg University, Denmark; both workshops were held at their premises in Copenhagen.

In addition to the authors of the contributions contained in this book, to whom we owe a considerable debt of gratitude, we would also like to warmly thank Professors Inge Røpke and Susse Georg for their contributions to the workshops in Copenhagen and, in different ways, to enabling this book. Inge was the principal investigator on the ECOMAC project and generously advised and supported us regarding administrative details. Chiara Fratini and Andrés Felipe Valderrama Pineda contributed both to some of our formative discussions on urban transitions and governance and, subsequently, through conference sessions and workshops in London and Copenhagen.

Many other individuals have influenced the development of the ideas expressed in the editors' contributions to this book, through many years of discussions and collaborations. Listing the people who influenced us can be ponderous for the reader so we have tried to be very selective. We would like to express particular thanks to the following individuals who have shaped us and/or the framing and content of this book: Tim Richardson, Jan-Peter Voss, Rob Raven, Harald Rohracher, Harriet Bulkeley, and Sheila Jasanoff.

Finally, as we write this preface we reflect on the pressing environmental challenges and gross social injustices that face urban populations. The United Nation's rallying cry of 'transforming the world', while arguably a somewhat shallow political commitment that is weakened by its association with the historical failures of the sustainability agenda, reflects a fundamental belief underpinning this book on the need for radical societal change. Lest we forget, the discussions about theories, hypotheses, analytical perspectives and case studies are set against a background of enabling experimental governance that, perhaps, can make a modest contribution to political, economic and social change.

Jens Stissing Jensen, Matthew Cashmore and Philipp Späth
Copenhagen, Denmark
Ås, Norway
Freiburg, Germany

1 Introduction

The knowledge politics of urban sustainability transitions

Matthew Cashmore,[1] *Jens Stissing Jensen*[2] *and Philipp Späth*[3]

Introduction

Urbanisation during the last several hundred years generates contradictory sentiments amongst different actors. On the one hand, it can be interpreted as a centrepiece of human achievement – a vibrant and profound statement of social, technical and economic progress. For other actors, global urbanisation is a principal cause of the Anthropocene – an ecological crisis precipitated, in no small part, by the resource-intensive and polluting metabolisms of human activities in cities. And yet others associate urbanisation primarily with the gross social inequalities between the 'haves' and the 'have nots', with political failures and injustices, and the hegemony of capitalism.

Attitudes about the successes and failures of urbanisation processes clearly vary, but there is widespread recognition that the cityscapes of the future need to be markedly different from those of today if they are to be sustainable and liveable. A variety of terms have been used to portray potential urban futures, reflecting different views on the role of technology (e.g. in innovative and smart cities), economic growth (e.g. green growth and de-growth) and ecology (e.g. in eco-cities or biophilic cities). Radical, systemic and coordinated urban change – an urban transition – is thus a cornerstone of discussions among policymakers, planning practitioners and academics alike.

In many respects, city administrations have been at the forefront of initiatives to promote more sustainable and liveable futures. The willingness of urban actors to play a leading role in the agenda on climate change is particularly noteworthy. Urban sustainability experiments – purposeful small-scale interventions that test alternative urban futures – have been occurring in considerable numbers across the globe. Nevertheless, there has been an overwhelming failure to convert the political goodwill embedded in experimental interventions into effective governance for, and of, systemic transitions (Bulkeley and Castán Broto 2013). One explanation for these apparent failures that is currently receiving increasing attention is that the political dimensions of transitions have been insufficiently conceptualised (Scoones *et al.* 2015).

Urban transitions are inherently and deeply political, and for a variety of reasons. They challenge existing relations of power and capital, involve trade-offs amongst competing visions of the future that may fundamentally affect equality, ethics and liberty in society, and portend new forms of democratic organisation and engagement. In this book we are interested in one particular component of the politics of transitions: epistemic (or knowledge) politics.

As Scoones *et al.* (2015, p. 4) note, '[t]here is a politics around knowledge production in debates about green transformations, turning both on what we think we know (consensus and uncertainties) and on who knows it (whose knowledge counts)'. We contend that contemporary urban 'political rationalities' (or ways of knowing, thinking and acting) are intrinsically connected to the specific knowledge-configuring activities employed in rendering urban systems as intelligible and governable concerns. We also contend that urban systems are not self-evident entities, with naturally defined boundaries, functions and sustainability challenges. The way in which urban wastewater is acted upon, for example, is linked to the techniques and devices used to construct it as a knowable material system, and the types of expertise that are deemed relevant to managing it (Jensen *et al.* 2016). Knowledge-configuring practices produce visibilities that foreground particular characteristics of urban systems and obscure others, and in so doing they make some actors and actions more relevant to the system's operation than others.

The development of transformative political rationalities thus often presupposes the invention of new ways of knowing urban systems. Understanding knowledge politics is not simply a question of bringing to critical attention the previously obscure political dimensions of transition governance, for it can be a force for directing change in positive directions. It opens up new understandings of the stickiness of the status quo, how unequal relations of power are (re-) accomplished, and how winners and losers are determined. This knowledge can be enabling and emancipatory, for it allows society to more fully comprehend the implications of alternate futures and navigate pathways to those futures.

A focus on knowledge politics is particularly germane as the 'big data' agenda sweeps through cities. Expanded capacity for collecting and analysing large quantities of data is ushering in a new era in urban knowledge politics. The 'Urban Operating Systems' of the future are a case par excellence of how the apparent neutrality of analytical devices black-box the political constitution of urbanisation (Chandler 2015; White 2016).

In this book, we outline the contours of a novel research agenda exploring how knowledge-configuring practices are implicated in, and modulate the politics of, urban transitions. We are interested in exploring and explaining how new ways of knowing may contribute to the governance of transitions through the inclusion of particular concerns, interests and values. The analyses contained in the subsequent chapters explore the political processes through which knowledge-configuring practices and strategies for representing the urban fabric are assembled and institutionalised. The authors also explore how the outcome of these processes shape the political rationalities that underpin discussions over urban futures.

Towards a knowledge-politics perspective on urban transition governance

We have argued that an understanding of knowledge politics is important to governing urban transitions, both as a source of mutability and as a mechanism for steering change along sustainable and just pathways. In this section of the chapter we describe the principal inputs to an analytical perspective on knowledge politics. We draw on insights from three main fields of scholarship in conceptualising knowledge politics: the transitions literature; governmentality studies; and, science and technology studies.

Transitions literature

The transitions literature emerged during the last 20 years from an amalgam of academic traditions, including innovation studies, technology and history (Shove and Walker 2007). Within this literature, transitions scholars have been understandably cautious about engaging with questions of governing change given a widespread view that any attempt to evoke change in complex systems is likely to have highly unpredictable outcomes. Elisabeth Shove (2010), nevertheless, claims that transitions scholarship has conceptually retooled the study of sustainability issues and created important new intellectual space in which to consider governance for sustainability. Recognition of this contribution is reflected in the increasing use of transitions scholarship in other academic fields, such as urban studies, and in policy decisions.

The transitions literature has sought to understand how radical change takes place in urban systems and infrastructures, such as energy, transport and water systems (Kemp *et al.* 1998; Geels 2002; Markard *et al.* 2012). Such systems and infrastructures have been conceptualised as socio-technical complexes formed over time as people, technology and institutions co-evolve. A socio-technical system is understood to be dynamically stable due to the reciprocities created through the co-evolution of system components (Geels 2002). These systems thus tend to change gradually, as opposed to undergoing abrupt transformations over a relatively short period of time. The study of historical transitions has enabled scholars to further their understanding of how and why technological innovations and external pressures, in some instances, have destabilised extant systems (Turnheim *et al.* 2015).

Studies of historical transitions have thus provided tantalising insights into sources of system mutability (e.g. Geels 2005). However, the role of novelty in terms of technical innovation has arguably been overemphasised in a tradition dominated by innovation analyses. Our point of departure for this book is that new ways of knowing and representing socio-technical systems can also play an influential role in transitions. We also reason that the consideration of knowledge politics is important in governing transitions in directions that are sustainable and just.

What then does the transition literature have to say on knowledge politics? An increasing number of transitions scholars have emphasised the political nature of epistemic specifications of urban systems. Shove and Walker (2007), for example, argued that, 'the process of abstracting the "it" in question – the policy, the goal, the system – from its historical and contemporary environment is not just a technical matter of analysis but a political, constructed, and potentially contested exercise in problem formulation'. Following a similar line of reasoning, Smith and Stirling (2007) suggest that societal systems are rendered visible as governable objects according to particular epistemic frames. Shifting from one epistemic frame to another is not a question of getting closer to or further away from an 'accurate' representation of a system; nor is it about eliminating uncertainties. Rather, it is the outcome of power struggles where actors manoeuvre to define the nature of urban reality according to their own priorities and interests.

> [u]ncertainties are transformed rather than reduced, since reductions under one framing merely pose more questions under alternative framings. Ambivalences prevail over how governance should best cut into and simplify the sheer complexities of socio-technical systems and dynamic relations with complex natural systems. Power relations are recognized to pervade the negotiation of governance constructs. Ever more accurate appraisal is unlikely to see the progressive evaporation of such power relations.
>
> (Smith and Stirling 2007, p. 367)

Given these assumptions, effective governance of societal systems is seen to involve politically infused frame-making. Recognition that interest-driven frame-making plays a constitutive role in transition governance has led to greater appreciation of the reciprocities between knowledge production and politics. Thus, some scholars have claimed that the vocabularies and concepts for making systems visible should be understood as basic ingredients for mobilising and configuring new arenas in which transitions governance can be organised and performed (Jørgensen 2012).

Governmentality

While the importance of knowledge politics to urban transitions is increasingly recognised by transition scholars, research on this issue remains fragmented and piecemeal. Furthermore, existing work is primarily conceptual; few scholars have empirically addressed the knowledge politics of urban sustainability transitions until very recently. It is to the governmentality literature, and its substantial body of empirically grounded analyses, that we turn for further inspiration.

Governmentality scholarship is inspired by a Foucauldian analysis of the reciprocity between knowledge and power in modern governance. It focuses on the conditions under which particular governance regimes come into being, are maintained and transformed (Dean 2010). Governmentality is based on an encompassing understanding of government and views it as calculated attempts to direct human behaviour (or 'to conduct conduct') towards particular ends. Government is viewed, in consequence, as ubiquitous, undertaken by various actors (not only 'the state' and its representatives), and involves the application of some form of calculative reasoning in efforts to structure actors' actions.

A central focus in governmentality studies is the role of knowledge in the fabrication of societal power dynamics. Knowledge is understood to be a product of discourses that are fundamentally neither true nor false. Governmentality studies are thus concerned, in part, with how knowledge-configuring practices are enrolled in purposeful attempts to conduct conduct. This focus reflects a belief that ways of knowing are intimately linked to, 'how the world is apprehended; the ways in which particular technologies and experts are constructed and deployed; and the formation of particular kinds of bodies and subjects' (Rutherford 2007, p. 293). Knowledge production, and an ability to affect what knowledge is viewed as legitimate and authoritative, are thus interpreted as being central to modern government.

> All government, we suggest, depends on a particular mode of 'representation': the elaboration of a language for depicting the domain in question that claims both to grasp the nature of that reality represented, and literally to represent it in a form amenable to political deliberation, argument and scheming.
>
> (Miller and Rose 1990, p. 6)

In presenting his governmentality perspective, Foucault emphasised the importance to modern government of statistical analyses that led to the emergence of 'the population' as a governable object. The production of statistical population data gained influence in the early nineteenth century (Scott 1988) and has had a pronounced effect on urban governance (Hacking 1982). A pivotal characteristic of this new category was that urban phenomena were made visible at a highly aggregated level (i.e. the population), rather than at the level of the individual or family. This meant that phenomena that appeared aleatory – and thus un-governable at the levels of the individual or family – were rendered visible as probabilistic patterns at the level of the population (Dillon 2007).

One salient effect of the invention of the population was a new conception of the health problems of the early industrial cities. Prior to this, only a vague understanding of the relation between urban health and living conditions existed. Health constituted a social problem in need of political intervention only once population statistics rendered it visible in the form of, among other things, mortality and morbidity rates (Gandy 2006). During the nineteenth and twentieth centuries, concern about the welfare of urban populations encouraged new conceptions of urban planning and development to emerge. Perhaps most significantly, the hydrological infrastructures of the city were reframed as public health concerns and an important mechanism for improving the productivity and longevity of the urban population (Gandy 2006). Traditional models of water provision, based on privately operated water-supply systems, were replaced by publicly operated, centralised systems. Infrastructural interventions were complemented by disciplinary measures promoting personal hygiene and regular exercise (Gandy 2006). This example illustrates how transformative political rationalities can be predicated on new ways of making the urban fabric known.

The governmentality literature offers three important contributions to an understanding of knowledge politics. First, it illustrates how knowledge plays a constitutive role in urban governance because it renders the urban fabric visible in particular, and inherently subjective, ways via the boundaries, purifications, differences and commonalities enacted under particular epistemic practices. By defining how urban reality is understood, knowledge production establishes the conditions under which political rationalities are formulated.

Second, the governmentality literature emphasises that discursive narratives, on their own, seldom constitute authoritative knowledge. Rather, the production of authoritative knowledge is typically rooted in the epistemic devices used to produce and configure knowledge. Accordingly, Miller and Rose (1990) argue that a governmentality inspired analysis needs to pay attention to:

> (…) the particular technical devices of writing, listing, numbering and computing that render a realm into discourse as a knowable, calculable and administrable object. 'Knowing' an object in such a way that it can be governed is more than a purely speculative activity: it requires the invention of procedures of notation, ways of collecting and presenting statistics, the transportation of these to centres where calculations and judgements can be made and so forth.

A contemporary example of how epistemic devices shape urban governance is the influence of so-called city league tables that rank cities on 'liveability' and 'business friendliness', amongst other things. High rankings are presented as a proxy for the attractiveness of a city, typically in terms of international capital and skilled labour. These analytical comparisons have enabled international inter-city competition to become an important political rationality in urban governance. Kornbeger and Carter (2010), in reflecting on the impact of city league tables argue that they function as 'engines' of competition rather than objective 'cameras'. They produce, 'classifications that define the boundaries of a field, identify players within that field, and relate them to each other. (…) In turn, these rankings produce a certain order of things that constitutes the a priori for competition' (Kornberger and Carter 2010, p. 328).

Third, the governmentality literature suggests that power relations in society are intimately related to knowledge and its authorisation. Political power is understood as a fundamentally productive and facilitative phenomenon; consequently, government is achieved not through direct prohibitions and controls, but by 'making-up' (after Hacking 1982) citizens who act in particular ways because they believe it is in their self-interest. Those actors who can influence how sociotechnical systems are understood can thus exercise political power, in theory at least, in subtle and rather opaque ways.

Science and Technology Studies

Science and Technology Studies (STS) is the third field of enquiry that informs our interest in knowledge politics. Some of the foundational contributions to this literature consist of ethnographic studies of scientific knowledge production. Echoing some of the findings from the governmentality literature, early studies in this tradition traced how scientific knowledge production made use of highly controlled devices and procedures in order to construct context-independent knowledge from complex real-world situations. In these studies, scientific knowledge production is portrayed as interdependent processes of 'simplification' and 'amplification' (Latour 1999). Simplification refers to processes involved in reducing complex situations into simple and context-independent knowledge objects: for example, by reducing a forest with thousands of plants and animals to a specific sample of leaves that can be transported to a laboratory to be categorised and further examined. 'Amplification' is achieved as all botanists use the same devices and procedures in cataloguing biodiversity. This enables knowledge produced in one forest to be readily compared and combined with knowledge produced in other forests. By rendering invisible the context specificity of complex real-world situations, through standardised simplification processes, knowledge objects that are readily comparable across time and space are created.

Similarly to the governmentality literature, STS emphasises that analytical knowledge should not be mistaken for neutral reflections of reality, but that knowledge production involves systematic reductions and often requires (standardised) material devices. What becomes visible and invisible as components of the real world and is translated into knowledge objects is thus contingent upon the particular epistemic devices used.

What sets the STS literature apart from the governmentality literature is its detailed attention to the multiplicity of ways in which knowledge objects are produced. Studies of professional-knowledge practices, for example, illustrate that analytical knowledge is seldom entirely consistent, and that closure over how to know a given situation is normally a transient occurrence. Building on the traditions of STS, governmentality studies and interpretive political sciences, Voß and Freeman (2016) have traced how knowledge on governance is co-produced together with (knowledge on) the objects that are expected to be governed. These studies demonstrate that even seemingly well-defined phenomena are sometimes rendered visible in divergent ways under different knowledge-production practices. For example, Mol (2002) shows that a rather everyday disease such as atherosclerosis is, in practice, enacted as partly different objects under different professional-knowledge practices. Another example of how closure on how to know a situation may be difficult to achieve concerns professional practices pertaining to the valuation of financial assets. In this case, analysts sometimes come up with very different valuations, because they construct and employ different 'valuation frames' to calculate the worth of an asset (Buenza and Garud 2007).

Callon (1998) employs the notions of 'framing' and 'overflowing' to conceptualise the practical difficulties of achieving closure over how to transform complex real-world situations into knowable and governable objects. In his terminology, framing refers to material and discursive networks involved in defining the boundaries of a situation by 'putting the outside world into brackets' (p. 249). Such material-discursive networks are portrayed as a necessary precondition for effective governance because they are instrumental in defining the relevant objectives of the situation and help to distribute roles among engaged actors and technologies. Framings may be subject to overflows, however, when the effects of actors, technologies or materials presumed irrelevant or under control, can no longer be foreseen, ignored or taken for granted. In situations where existing framings are compromised by such overflows, well-established facts, interests, identities, responsibilities, priorities and technologies all tend to become uncertain. A key point is that the objectives, identities and relations among actors require stabilised material-discursive framings of situations. In cases where material-discursive framings are somehow destabilised, Callon (1998) argues that:

> [...] the actor's ontology is variable: his or her objectives, interests, will and thus identity are caught up in a process of continual reconfiguration, a process that is intimately related to the constant reconfiguration of the network of interactions in which he or she is involved.
>
> (p. 253)

Material-discursive framing processes are thus perceived as deeply political, for actors' social, professional and political worlds are at stake when framings are compromised, along with the power relations and patterns of domination that define these worlds.

Inspired by this political perception of material-discursive framing processes, studies have investigated how new knowledge-producing devices may be employed to generate overflows that transform professional actors' established objectives and interests. For example, in a study of an urban development project, Georg (2015) explores how a mandatory accounting tool for urban sustainability influenced proposals for a new urban district developed by a series of interdisciplinary teams of architects and engineers. The study demonstrated that most teams applied the accounting tool as a largely passive 'intermediary', i.e. as a tool to evaluate their proposal after completion of the design processes. These teams thus attributed larger importance to traditional design criteria, such as place sensitivity. Some teams, however, allowed the accounting tool to more actively overflow their traditional framing of the situation. In these cases, the tool-based knowledge was used to continuously evaluate and shape design solutions. This study illustrates how multiple framing devices, which provide different – and sometime competing – ways of representing the order-of-things, may be involved in rendering complex situations governable.

A knowledge-politics perspective: fundamental premises

The three fields of scholarship – transitions studies, governmentality and science and technology studies – cumulatively provide the contours of a knowledge-politics perspective on transitions governance. We now review and summarise the fundamental principles of this perspective.

It has been repeatedly suggested that urban systems are not discrete phenomena 'out-there', but that such systems are rendered visible as objects through the application of knowledge-configuring practices. The politics of such practices lie, in part, in the concrete purifications and abstractions employed to translate unruly realities into governable objects. They also arise from the empowerment and disempowerment of actors and actions under these practices. The foundation of the epistemic politics approach is a constructivist belief that knowledge about urban systems is subjective, contingent, privileged and productive. From this point of departure, central concerns under a knowledge-politics perspective are:

- What knowledge-configuring practices are involved in rendering the urban fabric governable?
- What visibilities and invisibilities do these practices create?
- Who, or what, is empowered and who is disempowered by these practices?
- How do knowledge-configuring practices influence governance?

A knowledge-politics perspective thus draws attention to the knowledge-configuring practices employed to render the urban environment visible. Such practices take many forms: from disciplinary traditions in particular scientific and professional fields (e.g. the use of maps in geography), to calculative instruments that produce numerical representations and facilitate the making of comparisons (e.g. statistics). We view all knowledge production as inherently political, for it is fundamentally productive.

We use the term political rationality (after Dean 2010) to describe particular ways of thinking about, calculating and responding to problems. A political rationality is understood as the practical reasoning underpinning urban governance. Political rationalities are important because they determine, for example, the institutionally embedded ways of specifying what is problematic in a city and the methods of calculation that delimit how solutions to these problems are constructed. A premise of the knowledge-politics perspective is that ways of knowing, political rationalities and governance actions are intimately linked.

The knowledge-politics perspective is based on a premise that knowledge becomes analytical through processes of constructing boundaries, assigning functions and highlighting relations, while at the same time filtering out those complexities deemed irrelevant or unknowable. An important analytical function of knowledge production used in urban governance is to reduce the complexity of a system via the formulation of simplifications based on particular subjective assumptions. Analytical knowledge thus constitutes a purification of real-world complexities through political choices about what is relevant and what is not. In this way, knowledge is deeply entangled in the construction of political rationalities as knowledge-configuring practices enable the urban fabric to be defined. Analysing the knowledge politics of contemporary urban governance, therefore, requires attention to be given to how actors are empowered/disempowered under situated knowledge-production arrangements and their linked political rationalities. This entails sensitivity to how system challenges are understood: for example, which actors get to define the challenges and solutions; and how expertise and political authority is distributed among private and public actors.

A further premise of the knowledge-politics perspective is that transition governance will involve a transformation of established ways of knowing and intervening in socio-technical systems. While transitions may entail the introduction of innovative technologies, it can, and we argue must, encompass new ways of knowing and problematising established systems.

Sustainability and the politics of radical change

It is important to emphasise that politics is inherent to the notion of governance for sustainability. It is a politics that aims to re-calibrate the instrumentation of governance so as to make the urban fabric visible in the guise of social and ecological sustainability challenges; a politics that promotes radical change along novel, uncertain and contested pathways; and, a politics that transmogrifies existing societal power relations. Urban sustainability transitions, in short, is a topic replete with moral and ethical considerations.

Ever since the Rio Earth Summit of 1992 there has been considerable debate about what the notion of sustainability, its idioms and related terms mean and their utility as a framing of the issues at stake and of political goals. The apparent ambivalence of the term sustainable development, in particular, has been variously interpreted as a critical weakness that renders it essentially meaningless (Carruthers 2005) or alternatively as the genius of its inception as a metadiscourse bridging ostensibly irreconcilable world views (Meadowcroft 2011). Some commentators have suggested it is an intentionally nebulous construct in order that 'endless streams of academics and diplomats could spend many comfortable hours trying to define it without success' (O'Riodan 1993, p. 37). Yet others have seen a violence of sorts in its vagueness, noting that the term sustainability can be appropriated for multiple ends: for example, in legitimising the displacement of marginalised groups for what is claimed to be their material benefit or in supporting exploitative labour practices underpinning hegemonic models of economic growth. Swyngedouw (2009) attributes this to what he describes as the illusionary post-political common interest claims of sustainability that serve to mask vastly unequal power relations.

Although the heady days of vibrant academic, political and societal discussions on sustainable development of the 1990s have been eclipsed, the debate on enduring societal futures continues apace under various labels: the low-carbon economy, degrowth, the green economy, the circular economy, the bioeconomy, the smart economy and planetary boundaries, to name but a few. The terminology used to describe visions of the future clearly illustrate the centrality for some actors of questions concerning economic growth to these debates (Scoones *et al.* 2015). The Sustainable Development Goals (SDGs), adopted in 2015 under the auspices of the United Nations (UN), are a high level political statement that attempts to extend the lifetime of sustainable development and the legacy of the Rio Earth Summit. The UN's headline slogan that the SDGs will 'transform our world' emphasises the scale of changes encompassed within its headline ambitions of ending poverty, protecting the environment and ensuring prosperity for all. Yet the wisdom of attempting to reinvigorate a political agenda that for many actors has become little more than empty rhetoric has been questioned.

There are clearly multiple visions of what a sustainable future might look like: from the more technoscientific notions embedded in the idea of a new industrial revolution (e.g. European Commission 2018) to populist notions promoted through community, grassroot networks. These visions reflect competing framings of the problem and solutions underpinning the need for a sustainability transition and, as such, matter in very important ways to actors and interests. The contributors to this book do not claim to know precisely what sustainable urban futures look like, nor do they offer recipes for how socio-technical systems should change, although they may have opinions on desirable directions and the ways in which society attempts to articulate and move towards these futures. The primary focus of this book, instead, is on how and why knowledge matters politically to governance for sustainability. Within this context, sustainability is a useful concept for our endeavour in so far as it serves to express a desire for a radical renegotiation of human relations with the world around us. Yet it is also a term that we use with slight trepidation, given the aforementioned criticisms of the concept.

In focusing on knowledge politics, we are simultaneously drawing attention to certain aspects of the politics of sustainability transitions and black-boxing others. Whilst not intending to downplay the importance of other dimensions of the politics of sustainability transitions, our modest collective ambition in the publication of this book is to shed analytical light on one important component, which we have labelled knowledge politics.

Outline of the book

Drawing upon the premises of the knowledge-politics perspective we have outlined, the contributors to this book explore a variety of dimensions of knowledge politics and the dynamic interlinkages between power, knowledge and governance. Central to the collective contribution this edited volume makes are the stories it contains of knowledge experimentation, controversies and contestation, inequalities and asymmetries, and policy failures. These contributions cumulatively provide a rich exploration of knowledge politics and advance our understanding of the role of knowledge politics in governance for sustainability transitions.

In Chapter 2, Matthew Cashmore expands upon the conceptual terrain of the knowledge-politics perspective by examining the meanings of and interlinkages between politics, power and knowledge. He introduces the encompassing definition of politics that underpins our analytical endeavours in this volume and explores why knowledge matters politically using the New York High Line urban redevelopment project as a case study. The case study demonstrates how different ways of knowing the urban environment materially affect the types of urban futures that are conceivable, with potentially profound social consequences. The author also reviews the way in which the notion of political power has been conceptualised and analysed in transitions studies.

Chapter 3 and 4 shed light on the relationship between new ways of knowing and established urban governance arrangements. Recent debates on urban futures have been extensively influenced by discussion on the role of smart data and technologies in knowing and optimising the city (Klauser and Söderström 2014; Meijer and Bolívar 2016). In Chapter 3, Evelien de Hoop and colleagues discuss how a smart-city programme in Barcelona, and hence urban knowledge production, has been affected by shifting power relations among the political class. This shift occurred in 2015 when a market-oriented city leadership was replaced by one with origins in social movements. This produced a shift in the dominant political rationality informing the smart-city programme: from one that accommodated private sector control over smart infrastructure and data, based on a belief that they were best placed to effectively manage urban metabolisms, to one emphasising citizen self-determination and their control over smart data.

While the authors of Chapter 3 explore how the configuration of smart-city knowledge is modulated by power relations, Jens Stissing Jensen and Ulrik Jørgensen argue in Chapter 4 that new ways of knowing can also sometimes contribute to the reconstruction of political identities and interests. The analysis in this chapter illustrates, amongst other things, how urban planners in Copenhagen creatively exploit the expert-based vocabulary of socio-economic analysis to reframe cycling as a mode of urban mobility that has significant positive impacts upon public health. This way of knowing has become an important political rationality in the promotion of cycling. The authors of Chapter 4 show that hegemonic actors are not always in complete control of knowledge production and explore how power dynamics sometimes may be influenced by the introduction of new ways of knowing.

Chapters 5, 6 and 7 examine how politics is embedded in the 'nuts and bolts' of urban knowledge production. The author of Chapter 5, Nanja Christina Nagorny-Koring, uses a knowledge-politics perspective to analyse the application of 'best practices' as a salient knowledge device for the governance of low-carbon urban futures in Germany. The author argues that best practice is not a transparent epistemic tool that governance actors can use to orchestrate effective transition experiments, but rather it is a political technology that urban actors use to render particular technologies, political agendas and governance strategies visible. The author also explores how the concept of 'best practice' tends to obscure a coherent system perspective on low-carbon governance by making the urban fabric visible in the guise of discrete practices and technological solutions. The analysis traces the underlying power/knowledge complex of 'best practice' and advances our understanding of how the governance of low-carbon futures operates politically, on the ground.

Jens Stissing Jensen and Peter Karnøe in Chapter 6 analyse the knowledge devices and practices used to render the provision of heating intelligible as a governable phenomenon in Demark. Their analysis charts how the national government administration defines investment decisions using various price signals that are constructed on the basis of contentious configurations of situated analytical assumptions. The authors chart how urban governance actors have attempted to reconfigure the analytical assumption underpinning the heating system to promote coordinated investment in low-carbon technologies. These attempts, however, have been largely marginalised by the dominant expert community, which is located within the national government administration. The case indicates that a reconfiguration of the expert-based networks underpinning existing governance arrangements may be required in order to create opportunities for the development of knowledge for low-carbon governance.

In Chapter 7, Vanesa Castán Broto and Louise Guibrunet analyse the politics of a discrete knowledge device, material flows analysis, and examine its application in rendering the waste system in Mexico City knowable. The authors argue that this tool is instrumental in rendering important sustainability challenges, like informal waste scavenging practices, invisible and hence largely removing them from the political agenda. The case illustrates how knowing a system necessarily involves the definition of boundaries between the system and its context, and that the way in which these boundaries are defined has important political repercussions.

Chapters 8 and 9 report on urban experiments that were intended to restructure governance activities by configuring anew the ways of thinking about the urban environment. Matthew Cook and colleagues critically examine in Chapter 8 a governance process where they themselves were involved in configuring smart-city knowledge in the context of a multiplicity of urban stakeholders and interests. The analysis draws centrally on Flyvbjerg's (2001) conception of phronetic research. The smart-city experiment revolved around the development of a mobility app for the 'new town' of Milton Keynes, in the UK. The case illustrates the political sensitivities associated with the collection and configuration of data in and for the smart city. Retailers prevented data from being collected, for example, due to a fear that citizens might use this information to revise their shopping patterns, whereas a social movement argued that data production should be configured so as to enable them to hold public transport providers to account. The case illustrates how knowledge production in the smart city is a malleable phenomenon and exposes the power dynamics involved in an initiative to reorder the status quo.

In Chapter 9, Shivant Jhagroe and Derk Loorbach draw lessons from a case where researchers employed a set of epistemic techniques inspired by complex adaptive-system thinking to render the mobility system in Rotterdam visible in new ways. A visioning process orchestrated using these epistemic techniques problematised the conventional governance objectives, which centred on safety and the speed of cars, and rendered mobility inequities among the various parts of the urban population visible as a governable concern. The case illustrates how deliberately organised knowledge-configuring processes can help to link systems governance to experimental urban agendas that enable society to chart and explore alternative mobility pathways.

In Chapter 10, we conclude by summarising the key contributions this book has made to understanding – conceptually and empirically – the knowledge politics of sustainability transitions. We argue that cumulatively the contributions vividly highlight the importance of the apparently mundane devices configuring everyday knowledge production. It is suggested that alternate ways of knowing have the potential to make new pathways to sustainability visible and challenge the power relations underpinning political rationalities that support unsustainable practices.

Notes

1 Department of Urban and Regional Planning, Norwegian University of Life Sciences, Norway.
2 Department of Planning, Aalborg University, Denmark.
3 Institute of Environmental Social Sciences and Geography, University of Freiburg, Germany.

References

Buenza, D. and Garud, R., 2007. Calculators, lemmings or frame-makers? The intermediary role of securities analysts. *The Sociological Review*, 55 (2), 13–39.

Bulkeley, H. and Castán Broto, V., 2013. Government by experiment? Global cities and the governing of climate change. *Transactions of the Institute of British Geographers*, 38 (3), 361–375.

Callon, M., 1998. An essay on framing and overflowing: economic externalities revisited by sociology. *The Sociological Review*, 46 (1), 244–269.

Carruthers, D., 2005. From opposition to orthodoxy: the remaking of sustainable development. In J.S. Dryzek and D. Schlosberg, eds. *Debating the Earth: The environmental politics reader*. Oxford: Oxford University Press, pp. 93–112.

Chandler, D., 2015. A world without causation: big data and the coming of age of posthumanism. *Millennium: Journal of International Studies*, 43 (3), 833–851.

Dean, M., 2010. *Governmentality: Power and rule in modern society*. London: Sage.

Dillon, M., 2007. Governing through contingency : the security of biopolitical governance. *Political Geography*, 26, 41–47.

European Commission, 2018. *Capitalising on the benefits of the 4th industrial revolution*. Luxembourg: Publications Office of the European Union.

Flyvbjerg, B., 2001. *Making social science matter: Why social inquiry fails and how it can succeed again*. Cambridge: Cambridge University Press.

Gandy, M., 2006. The bacteriological city and its discontents. *Historical Geography*, 34, 14–25.

Geels, F.W., 2002. Technological transitions as evolutionary reconfiguration processes: a multi-level perspective and a case-study. *Research Policy*, 31 (8–9), 1257–1274.

Geels, F.W., 2005. The dynamics of transitions in socio-technical systems: a multi-level analysis of the transition pathway from horse-drawn carriages to automobiles (1860–1930). *Technology Analysis & Strategic Management*, 17 (4), 445–476.

Georg, S., 2015. Building sustainable cities: tools for developing new building practices? *Global Networks*, 15 (3), 325–342.

Hacking, I., 1982. Biopower and the avalanche of printed munbers. *Humanities in Society*, 5 (3–4), 279–295.

Jensen, J.S., Fratini, C.F. and Cashmore, M.A., 2016. Socio-technical systems as place-specific matters of concern: the role of urban governance in the transition of the wastewater system in Denmark. *Journal of Environmental Policy & Planning*, 18 (2), 234–252.

Jørgensen, U., 2012. Mapping and navigating transitions – the multi-level perspective compared with arenas of development. *Research Policy*, 41 (6), 996–1010.

Kemp, R., Schot, J. and Hoogma, R., 1998. Regime shifts to sustainability through processes of niche formation: the approach of strategic niche management. *Technology Analysis & Strategic Management*, 105 (2), 175–198.

Klauser, F. and Söderström, O., 2014. Michel Foucault and the smart city: power dynamics inherent in contemporary governing through code. *Environment and Planning D: Society and Space*, 32 (2013), 869–885.

Kornberger, M. and Carter, C., 2010. Manufacturing competition: how accounting practices shape strategy making in cities. *Accounting, Auditing & Accountability Journal*, 23 (3), 325–349.

Latour, B., 1999. Circulating reference: sampling the soil in the Amazon forest. In B. Latour, ed. *Pandora's hope: Essays on the reality of science studies*. Harvard, MA: Harvard University Press, pp. 24–79.

Markard, J., Raven, R. and Truffer, B., 2012. Sustainability transitions: an emerging field of research and its prospects. *Research Policy*, 41 (6), 955–967.

Meadowcroft, J., 2011. Engaging with the politics of sustainability transitions. *Environmental Innovation and Societal Transitions*, 1 (1), 70–75.

Meijer, A. and Bolívar, M.P.R., 2016. Governing the smart city: a review of the literature on smart urban governance. *International Review of Administrative Sciences*, 82 (2), 392–408.

Miller, P. and Rose, N., 1990. Governing economic life. *Economy and Society*, 19 (1), 1–31.

Mol, A., 2002. *The body multiple: Ontology in medical practice*. London: Duke Universtity Press.

O'Riodan, T., 1993. The politics of sustainability. In R.K. Turner, ed. *Sustainable environmental economics and management: Principles and practice*. Chicester: John Wiley & Sons, pp. 37–69.

Rutherford, S., 2007. Green governmentality: insights and opportunities in the study of nature's rule. *Progress in Human Geography*, 31 (3), 291–307.

Scoones, I., Newell, P. and Leach, M., 2015. The politics of green transformations. In I. Scoones, M. Leach and P. Newell, eds. *The politics of green transformations*. Abingdon: Routledge, pp. 1–24.

Scott, J.C., 1988. *Seeing like a state: How certain schemes to improve the human condition have failed*. London: Yale University Press.

Shove, E., 2010. Social theory and climate change. *Theory, Culture & Society*, 27 (2–3), 277–288.

Shove, E. and Walker, G., 2007. CAUTION! Transitions ahead: politics, practice, and sustainable transition management. *Environment and Planning* A, 39 (1998), 763–770.
Smith, A. and Stirling, A., 2007. Moving outside or inside? Objectification and reflexivity in the governance of socio-technical systems. *Journal of Environmental Policy & Planning*, 9 (3–4), 351–373.
Swyngedouw, E., 2009. The antinomies of the postpolitical city: in search of a democratic politics of environmental production. *International Journal of Urban and Regional Research*, 33 (3), 601–620.
Turnheim, B., Berkhout, F., Geels, F., Hof, A., McMeekin, A., Nykvist, B. and van Vuuren, D., 2015. Evaluating sustainability transitions pathways: bridging analytical approaches to address governance challenges. *Global Environmental Change*, 35, 239–253.
Voß, J.-P. and Freeman, R., 2016. *Knowing governance: The epistemic construction of political order*. New York: Palgrave Macmillan.
White, J.M., 2016. Anticipatory logics of the smart city's global imaginary. *Urban Geography*, 37 (4), 572–589.

2 Governing radical societal change

Politics, power and knowledge

Matthew Cashmore

Introduction

A foundational belief of transitions scholarship is that radical change, in the form of transitions in coupled sociotechnical systems, are a requirement for more sustainable and just urban futures (Raven *et al.* 2016). This places questions of governance (defined in this chapter as the purposeful steering of transformative[1] experiments) at the heart of transition studies. The contribution that transition scholarship has made to retooling the way in which we think about governance for sustainable and just futures is increasingly widely appreciated in both academic and policy circles (Shove 2010; Scoones *et al.* 2015). Yet the lack of attention historically given to the politics of transitions is remarkable for a field of endeavour engaged centrally with questions of radical and systemic socio-technical change. Criticisms of the political naivety of transitions scholarship are now well rehearsed and a growing number of contributions, from various academic disciplines, have sought to redress this important lacuna.

Politics is often equated to activities associated with the administration of sovereign states or what can be described as the macropolitics of nationhood (Jasanoff 2004). In the political sciences, however, the term tends to be interpreted considerably more broadly. The influential political scientist Harold Laswell suggested politics essentially constitutes the struggle over who gets what, when and how (Laswell 1935). This definition implies that politics is not limited solely to the acts of sovereign governments and their administrations, nor is it purely concerned with the study of law. Politics is understood by Laswell as a phenomenon occurring in a multiplicity of arenas at international, national and local levels as actors seek to exert an influence on decisions concerning 'who gets what'. A focus on influence also foregrounds questions of power and, more specifically, how and why some actors are able to exert an influence, while others may appear relatively powerless.

Based on a Laswellian understanding of politics, I explore in this chapter a number of concepts and premises central to this book. I start by exemplifying the assertion made in Chapter 1 that knowledge matters politically in the governance of transitions, using the New York High Line Park as a case study. I then turn to consider a concept that is at the heart of political science and the knowledge-politics perspective – power. I introduce the multiplicity of ways in which power can be understood and examine how questions of power have been addressed to date in transitions scholarship. The chapter concludes with a brief review of the implications of the analysis contained in this chapter for how we understand, conceptualise and seek to govern sustainability transitions.

Why knowledge matters politically: the case of the New York High Line Park

The particularities of urban politics, and the shifts in political authority that they portend, is a subject that has been increasingly emphasised in recent years (Bulkeley *et al.* 2015). In this section I consider an example of these particularities which I use to exemplify how and why knowledge matters politically. The case study employed is an emblematic example of the repurposing of industrial urban infrastructure, the New York High Line Park, which first opened to the public in 2009. Since this time, the High Line Park has been lavishly praised as an innovative experiment, involving new models of public–private collaboration. The influential critic Paul Goldberger described the park as, 'the most important new model for public space in our time' (Green 2013). On the opening of the third and final phase of the park in 2014, the *New York Times*' architecture critic stated that '[n]ot since Central Park opened in 1857 has a park reshaped New Yorkers' thinking about public space and the city more profoundly' (Kimmelman 2014). The park also won numerous awards and quickly became established as a staple tourist attraction for visitors to New York City.

Although only a microcosm of an urban socio-technical system, the High Line Park is an example of urban redevelopment that is replete with questions concerning winners, losers, sustainability and justice. The High Line originally was an elevated urban railway that ran through parts of west Manhattan. The railway had lain unused since the 1980s and it was widely viewed as an unwelcome reminder of urban decay and an obstacle to inward investment. Sections of the railway were dismantled in 1960 and 1991, and there was strong support amongst local property owners and the city administration for dismantling the remaining sections (Lang and Rothenberg 2017). However, opinion started to change around the end of the twentieth century and an upswell in support for its repurposing led to a decision being made to convert the remaining sections of the High Line into an urban park or promenade.

A non-governmental organisation, the Friends of the High Line (FHL), which was formed in 1999, was instrumental in mobilising support for the preservation of the remains of the High Line. The founders of the FHL commissioned the photographer Joel Sternfeld to document the derelict High Line. Sternfeld's photography drew attention to the dramatic juxtaposition of the emergent ecology that had grown on the High Line with the surrounding built environment. These pictures – knowledge rendered as art – were instrumental in generating support amongst the societal elite for preserving the High Line. It was, for example, Michael Bloomberg – the former Mayor of New York City –

who wrote an introductory essay for a book of Sternfeld's photography, *Walking the High Line*, which was published in 2001. Furthermore, the visual images played to the aspirations of the cultural elite, particularly of art gallery owners who had started to move in to the area.

The primacy of knowledge rendered as art was subsequently challenged by the then Deputy Mayor for Economic Development in Bloomberg's administration. Although purportedly sympathetic to the proposal for a park, the Deputy Mayor is reported to have reacted in the following way when presented with Sternfeld's photography at a meeting with the FHL:

> Don't show me pretty pictures. We have so many parks already that we can't afford ... You're not telling me about the money. I need to know what this means financially.
>
> (David and Hammond 2011)

To address the rationality of the Deputy Mayor, whose support was required in order to secure public funding for the project, the FHL focused on configuring knowledge through monetary calculations and a cost-benefit rationality. A report was commissioned in which the projected financial benefits of the park were estimated to greatly outweigh the redevelopment costs. Financial benefits were calculated to accrue, in large part, from the redevelopment of surrounding neighbourhoods, and the incremental tax revenues that this would generate (David and Hammond 2011). The initial visibility created for the repurposing of the High Line, an aesthetic representation centred upon the preservation of industrial heritage, was subordinated. The High Line Park was remade as a locus for stimulating high-end urban redevelopment. The park has proven highly successful in this regard, with almost two billion dollars of private investment in the area being attributed to its existence some two years after it opened (McGeehan 2011).

The importance of economic calculations to the Deputy Mayor focused attention on design questions in new ways, for the High Line Park was interpreted as needing to provide certain design qualities if it were to stimulate large-scale redevelopment. The celebrated successional vegetation that had established on the High Line, and was popularised in Sternfeld's photography, was largely replaced in the first two sections of the park. Vegetative plantings were instead imagined as style and setting, rather than urban nature (Reichl 2016). The third section of the park employs an ecological aesthetic reminiscent of that captured in Sternfeld's photographs, but it is a highly managed and manicured aesthetic. Activities deemed discordant with the design-quality goals are prohibited in the park. This includes many activities that are normally associated with parks, including dog walking, cycling, barbequing and the playing of ball games. A variety of surveillance and disciplinary techniques are used to regulate activities and behaviours in the park (Cataldi *et al.* 2011; Reichl 2016).

The High Line Park is a project replete with political consequences in the form of winners and losers. Commentators have argued that the recreational needs of low- and middle-income local residents have been subordinated to the needs of the tourist industry and the cultural elite. Local community members have suggested that the park has little relevance to their lifestyles or needs (Lang and Rothenberg 2017). The differing appeal of the High Line Park to residents of New York is evident in some of the visitor data. Surveys indicate that the ethnic profile of visitors is greatly out of step with the diversity of the local and city-wide population (Reichl 2016), while queer spaces in the vicinity of the park have been displaced, in part purposefully through more aggressive policing of these spaces (Cataldi *et al.* 2011).

Gentrification was already under way in the areas that the High Line Park traverses before it was repurposed, but the contribution of the park to gentrification has been significant (Millington 2015). As a consequence of gentrification, Chelsea, the main neighbourhood in which the High Line Park is located, now has one of highest levels of income inequality in New York City (Navarro 2015). The changing population characteristics and the characteristics of the visitors to the park has also promoted commercial gentrification. Small businesses catering to local residents' needs and budgets have been replaced by ones focusing on luxury consumption of high-end food, fashion and culture (Lang and Rothenberg 2017). As Leach (2015, p. 37) concludes, 'knowledge politics matter because they are so intimately entwined with material political economy'.

The term epistemic injustice has been used to draw attention to how the application of knowledge in public policy affects the influence that different actors can exert. Ottinger (2017) defines hermeneutic injustice as a specific type of epistemic injustice pertaining to inequalities in epistemic resources underpinning an actor's capacity to engage in policy discussions. Dominant actors, which in this example I have labelled broadly as the political and cultural elite, typically have greater access to epistemic resources than marginalised groups. This certainly appears to be true of the FHL, which while oftentimes described as a community grassroots initiative, was a highly exclusive organisation.

Knowledges and their role in creating particular situated visibilities are not the only factors implicated in the creation of winners and losers from the repurposing of the High Line. Nevertheless, in seeking to create appeal for the project amongst the cultural and political elite, the FHL contributed to a politics of the High Line Park that could have been otherwise. What kind of development may have emerged, for example, if the dominant political rationality had foregrounded the health and well-being of the local community?

Power and its expression

The case study of the High Line Park serves to illustrate some of the important political dimensions of knowledge and of the real-world implications of knowledge politics for sustainable and just urban futures. A central concept in the political sciences in understanding such political phenomena is that of power; indeed, for some political scientists, the study of politics *is* the study of power

(Jordan and O'Riordan 2001). Transitions scholars have also recognised the centrality of power to the governance of radical systemic change, for transitions are widely understood as necessarily involving a shift in extant power relations (Avelino *et al.* 2016; Temper *et al.* 2018). Any discussion of the knowledge politics of sustainability transitions thus axiomatically invokes questions of power.

And yet power is an amorphous concept that is understood in multiple ways. The existence of divergent interpretations of power is partly a consequence of it having been employed for different analytical and normative purposes. How power is understood is also shaped by the way in which it is differentiated from linked concepts (e.g. authority, legitimacy, empowerment and freedom) in particular analytical and theoretical frameworks.

In this section of the chapter, I briefly introduce some of the principal ways in which power is understood. This is important in setting out the theoretical terrain covered in this book because the way in which power is understood can have an important influence upon prescriptions for governing sustainability transitions (Partzsch 2017a). I use as an analytical heuristic the following categorisation of conceptualisations of power to discuss the ways in which it is understood: power 'to'; power 'over'; power as coercion; power as consensus; and, power as a productive force (after Haugaard 2002). I then draw on this heuristic in examining tendencies in the ways in which power has been conceptualised by transitions scholars.

Power 'to' and power 'over'

An oft-used analytical generalisation in scholarship on power is to differentiate between two forms of power: power 'to' and power 'over'. Power 'to' can be understood as a generalised ability to get things done (after Parsons 1967). Under this conception, power is viewed as neither inherently positive nor negative; it is merely a phenomenon that is essential for action. The ability of actors to exert power 'to' is central to the modalities of realising change for many transitions scholars (e.g. Partzsch 2017b; Avelino 2017) and is often linked to the concept of human agency. This is partly a function of the dominance of particular theoretical perspectives in the social sciences and transitions scholarship (see the subsequent section on power and transitions scholarship), but it also reflects a strongly held normative belief for many transition, and other, scholars that individuals are capable of making a difference (e.g. Partzsch 2017b). The power 'to' of an actor is said to be structured by their access to capacities (e.g. skills and training) and resources (e.g. money and time). Thus, power 'to' is unequally distributed across society.

In contrast to the ability-based definition of power 'to',[2] power 'over' is understood as a relational construct, wherein one actor influences the actions of another (Pansardi 2012). For some scholars, power 'over' is neither automatically good nor bad. It may reflect hierarchical relations that are intended to benefit a group, such as deferral to the coach in decisions on team tactics. Yet other scholars interpret power 'over' in a more restricted manner as the 'probability that one actor within a social relationship will be in a position to carry out his own will despite resistance' (Weber 1978, p. 53). Thus, for Weber, power 'over' involves a reciprocal relationship of domination and subordination. Domination may thus be interpreted as either a particular form or expression of power 'over' or analogous to power 'over' (Allen 1999).

Power 'over' is expressed through a variety of mechanisms. First, it may involve one actor directly determining the actions of another. For example, a municipal government may institutionalise policies that not all city residents agree with, due to the political authority afforded to the municipal government under democratic rules. Second, it may accrue from the ability of a particular actor to set the agenda and thereby control what issues are, and are not, open to debate. The ability to exclude certain issues from a discussion may be a function of, amongst other things, invoking selective precedents, recourse to particular types of expertise, or a function of an actors' resources (e.g. the capacity to participate in meetings, etc.). Third, power 'over' may be exerted by influencing the thoughts and intentions of actors according to specific conventions. These conventions may take the form of, amongst other things, organisational rules and norms, traditions, religious beliefs, and scientific protocols. Such expressions of power were labelled as 'false consciousness' by Lukes (1974), for while actors may believe they are acting autonomously, they are actually reifying existing norms.

Power as coercion versus power as consensus

A second common point of debate concerning power pertains to whether it is expressed primarily through coercion or is (or at least normatively should be) consensual. Coercion operates by making non-compliance with a demand costly in some way. This is typically achieved through some form of threat – for example, of withholding something (love, finance or food) or physical restraint – and through recourse to violence (Haugaard 2006). Coercive power has received considerable scholarly attention due to the role that military conquests have played historically in the emergence of the nation state, and colonial and global powers.

A frequent point of discussion in scholarship on power has been the effectiveness of coercive power. Mills (1956, p. 171) famously suggested that 'the ultimate kind of power is violence' and the history of military conquest has been cited as evidence that, at least in the short term, coercive power can be highly effective. In contrast, Mann (1986, p. 26) describes violence as, 'the most concentrated, if bluntest, instrument of power', contrasting the rapidity with which geopolitical changes can occur as military forces emerge and wither with, amongst other things, the limited geographical reach and difficulties of enforcing authoritarian rule based on coercion. Violence may thus be interpreted as a comparatively weak expression of power, for it becomes necessary to employ it where coercive threats are ineffectual (i.e. where power is disputed). Arendt (1970) goes further in suggesting that while power is the foundation of all government, authority maintained solely through violent coercion is ultimately untenable.

Haugaard (2006, p. 64) opposes what he regards as a 'common and mistaken view that hegemonic power is based upon coercion', arguing instead that social order is primarily based on consensual systems of rules. Consensual rule is based on a belief that it is only legitimate to exercise power where society has given its consent. This does not necessarily mean that all individuals in a society need to agree upon every action nor that they automatically strongly support every action that they consent to. Rather, consensual rule typically operates through reliance on shared norms, values and beliefs. The embeddedness of such rules within society (for example, in culture) is said to be the foundation of social order by some scholars (Haugaard 2003).

Dialogue has a central role in many philosophies of consensual rule. One of the most well-known of these is Jürgen Habermas's theory of communicative action (Habermas 1984; Habermas 1987). Yet the potential for achieving a true consensus, unfettered by relations of power, has been questioned. It has been argued, for example, that practical constraints and tensions over societal goals inevitably lead to compromises being made in consensual ideals during dialogical processes (Connelly and Richardson 2004). For this reason, Haugaard (2006) suggests that most social interactions occur somewhere on a spectrum between the two extremes of coercion and consensus.

Power as a productive force

A third way in which power is understood, and an interpretation that is central to much of the writing in this book, is as an omnipresent social dynamic that is constitutive of our subjectivities and hence our realities. This understanding is closely linked to the work of Michel Foucault. Drawing upon studies of the operation of modern institutions of government, such as prisons, Foucault distinguished power from other types of societal phenomenon involving force, such as domination. Foucault viewed power as a creative and productive phenomenon that is not possessed by 'powerful' individuals, but is expressed in the societal actions, and inactions, of ostensibly autonomous individuals.

According to Foucault (2002), power is neither inherently good nor inherently bad. It is a social dynamic that is both enabling and impeding, revealing and obscuring. The particular way in which power produces reality is by structuring those actions that are conceivable and legitimate to societal actors according to particular norms, aspirations and beliefs. This perspective on power problematises the supposed binary between those that govern and those who are governed:

> As an immanent force which constitutes its own organization, not one imposed from above or from the outside, power is seen as coextensive with its field of operation. Power is practised before it is possessed and it is this that gives rise to the roundaboutness of power, not some facile notion that it is a shadowy force lurking in the murky recesses.
>
> (Allen 2003, p. 9)

It also brings in to focus the potential that human behaviour can be regulated and controlled through the deployment of strategies to conduct conducts by mobilising particular subjectivities, such as one that is premised on a belief that neoliberal economic principles should determine urban development trends.

Foucault argued that there was a co-constitutive relationship between power and knowledge. Power dynamics influence what knowledge is considered truthful and authoritative (e.g. what counts as science and the value of scientific versus, for example, indigenousness knowledges), while systems of meaning entail specific ways of acting (e.g. the types of behaviours required to produce scientific knowledge) (Haugaard 2002). This resonates with the notion of co-production in science and technology studies, wherein the way in which reality is understood is said to be inseparable from the way a society chooses to live. The socio-technical imaginaries present in society (after Jasanoff 2015), which shape our visions of the present and of desirable futures, can thus be understood as a product of this power/knowledge dialectic (Latour 1983).

Power and transitions scholarship

How have questions of power been addressed within those parts of the transitions literature that have engaged explicitly with this issue? My intention is not to systematically review the literature, but merely to present some of the tendencies evident in conceptualisations of power and to explore points of discussion concerning the interlinkages between power, knowledge and radical change. In so doing, I focus on scholarship that has dealt explicitly with power, although I note that there is commentary of relevance in scholarship that addresses power implicitly or in other terms (e.g. in terms of democracy or participation).

Much of the research within the transitions field that has engaged with questions of power has done so within the framework of the Multi-Level Perspective (MLP), which has been a cornerstone of analytical and conceptual work in transition studies since it was proposed by Frank Geels in the early 2000s (see Geels 2002; Geels 2004). A central feature of this framework is the delineation of three analytical (as opposed to spatial or hierarchical) levels. The three levels are: (1) niches, where radical social and technological innovations are said to arise; (2) socio-technical regimes, which are the locus of the institutions and practices that are said to structure the system under consideration; and (3) an exogenous landscape which influences the sociotechnical regime, but over which regime actors have little influence. A transition is said to occur under this framework as a consequence of the emergence of novel alignments of processes within and between each of these three levels.

A transition constitutes more than simply the adoption of a new technology originating from a niche, for it necessarily encompasses wider scale, systemic changes, such as alterations in social preferences and behaviours, new market arrangements, and new organisational or cultural rules (Geels *et al.* 2017). Nevertheless, transitions scholars have focused extensively on innovation, and typically technological innovation, as a primary impetus for the emergence of new alignments across the three analytical levels in the MLP. As such, there has been a strong focus on the power 'to' of individuals, to create and market innovations. This focus is oftentimes expressed in terms of an interest in the agency of actors, who may be operating on their own or in coalitions. The power 'to' of entrepreneurial, but subordinate, agents is often pitted against the dominance of a socio-technical regime and its incumbent actors (Hess 2016; Haan and Rotmans 2018). As a consequence, an important theme in scholarship addressing power has been on how to tilt the balance of power away from regime actors. A similar focus is also evident in much of the literature that addresses power and transitions outside of the MLP (O'Brien 2012), but the concepts used differ to greater or lesser extents.

The presumed dominance of a regime over niche actors (under the MLP) or of an elite within society has been explained in various ways. For example, regime actors are assumed to have access to more resources than other actors, from which they can harness more power 'to'. Regime actors may also exert power 'over' other actors for they are typically in a strong position to define what is on the policy agenda and what is left off. Incumbent business interests may coerce politicians – tacitly or explicitly – due to their importance to the political economy (Geels 2014).

Avelino and colleagues (Avelino and Rotmans 2011; Avelino 2017) draw on the notion of power as a capacity (i.e. as power 'to', an attribute of human agency) in seeking to conceptualise under the MLP how extant regimes can be destabilised. Their approach is based on a belief that everything that needs to be said about power can be expressed in terms of the capacity of individuals to achieve particular outcomes. This does not mean that it ignores power 'over', they argue, but that power 'over' is conceptualised as an ability of individuals (Avelino and Rotmans 2011). They delineate three mechanisms through which power is expressed. 'Reinforcive power' is described as the capacity of actors to reproduce the status quo; 'innovative power' is the ability of actors to create new resources (e.g. a new invention or a new local currency) that serve to lessen reliance on the extant regime; and, 'transformative power' is the capacity to foster structural change through the creation of new rules and norms (i.e. to alter the structures that underpin the status quo).

Avelino and colleagues' understanding of power as a human capacity leads them to foreground questions concerning the relative distribution of power within society and how transitions do, can or should empower and disempower particular actors (Avelino and Wittmayer 2016). Indeed, the notion of empowerment has generated increasing attention in transition scholarship, reflecting, in part, interest in the power asymmetries underpinning relations of domination and subordination. Various definitions of empowerment are employed. Partzsch (2017a), for instance, states that empowerment can be understood as a kind of power 'with' where it involves multiple actors working collectively. She suggests that collective empowerment is generated through a process that is designed for the common good of all actors, through the development of shared values, by finding common ground, and by generating collective strength. Learning is central to collective empowerment for it is through learning that actors begin to question received wisdom and extant norms, and develop new understandings. Partzsch (2017a) views the empowerment of individuals as something different: an outcome of power 'to'. Irrespective of the precise way in which empowerment is understood, what is central to the concept is the notion that empowerment allows individuals or collectives to positively shape their futures (Ahlborg 2017).

Temper *et al.* (2018) also assert that transitions necessarily involve destabilising the power relations underpinning patterns of domination and subordination:

> For us, radical transformation implies one which refers to a transformation of power structures and relations, from a situation of domination, injustice and violence and unsustainability to one of reduced violence, increased equality and flourishing. It entails challenging the sources of domination and oppression including capitalism, patriarchy, state-centrism and inequities along lines of race, caste, ethnic, gender, ableism, sexuality and others.
> (Temper *et al.* 2018, p. 752)

They focus on the role of social movements as a destabilising force, arguing, with justification, that social movements are a transformative mechanism that has been under examined in the transitions literature. Opposition and resistance by social movements to hegemonic regimes are understood as key mechanisms through which sustainable futures can be envisaged. Conflict is thus interpreted as a potentially productive force because it necessarily involves articulating criticisms of the status quo. The importance of agonistic and unruly politics is also emphasised by Stirling (2015). Indeed, Stirling argues that unruly mobilisations are more likely to support radical and open-ended transition than more controlled attempts at introducing new technologies and practices.

Geels (2014) has criticised the focus in transitions scholarship on mechanisms for destabilising the status quo. He argues that the conceptual language of inertia, path dependency and lock-ins in descriptions of socio-technical regimes has been employed in such a way as to obfuscate the political work involved in maintaining the status quo. What is also important, he argues, is explaining the stickiness of incumbent socio-technical regimes (and see also Hess 2016). Regime stability is thus interpreted by Geels as an accomplishment that must be continuously achieved, even where a socio-technical system appears essentially unchanged over time. This observation also calls into question precisely how automatised (Avelino and Rotmans 2011) the embeddedness of a regime or dominant actors are in reality.

This is not to say that regime actors axiomatically use their position to resist change. It has been observed that the Crown Estate (a regime actor with overall responsibility for managing the seabed in coastal waters in England, Wales and Northern Ireland) was instrumental in orchestrating the emergence of a value chain for an offshore wind energy industry in the UK (Raven *et al.* 2016). The support proffered by the Crown Estate appears to be a function of its mandate to develop financial revenue from the assets it owns and manages (the seabed), which has been operationalised through what the organisation describes as a strategy of 'conscious commercialism' (the Crown Estate 2018). It was also possible for the Crown Estate and other actors at the time to align a narrative on offshore wind energy utilisation with socio-political demands for progress towards domestic renewable energy generation targets.

A number of individuals have cautioned against interpreting transitions solely in terms of overcoming relations of domination (Haan and Rotmans 2018). Hoffman (2013), for example, explores the apparent contradiction concerning how niche actors may destabilise a socio-technical regime and open up alternative futures, whilst simultaneously operating within the boundaries of a system structured by those same rules. Hoffman draws on a relational understanding of power in suggesting that '[power is] constituted by the ways in which actors, artifacts, and resources accrue meaning in relation to one-another and is thereby not placed with either the individual or the collective but the relations that constitute both' (Hoffman 2013, p. 265). The example of how wind energy utilisation in Denmark was linked to various exogenous trends (e.g. a desire to restrict fuel imports) is used to illustrate the capacity of actors to creatively construct new and potentially destabilising socio-technical imaginaries. In so doing, Danish wind energy advocates were able to transform what was frequently interpreted as a key limitation of wind energy (a small-scale production technology and a distributed energy source) into a persuasive argument for supporting their imaginary. The insurgent wind-energy actors also generated relational power that led to systemic change, Hoffman (2013) concludes, by reinterpreting their practices and by forming alliances and new organisations.

Based on such observations, Hoffman and Loeber (2016) argue for a modified approach to understanding the relation between niches and regimes in the MLP on the basis of a claim that there is not a clear division between the two levels and that change occurs as a result of their dynamic interlinkages. They also highlight the importance of the malleability of knowledge about an artefact's meaning to transitions, suggesting that their empirical work indicates that innovations in the way artefacts are known are arguably more significant in explaining transition dynamics than technological or social innovations. Thus, Hoffman and Loeber (2016) reach a similar conclusion to the point of departure for this book, albeit from a different starting point: struggles over defining the way in which a system is known are central to the politics of sustainability transitions and an important mechanism for expressing power.

Relatively few scholars have engaged systematically with what I have categorised as a productive interpretation of power. An interesting exception is the work of Ahlborg (2017), who combines a focus on agency as a relational phenomenon with a productive perspective on power to draw attention to the co-development of human relations and infrastructure: '[p]ower relations are produced through situated and embodied practices that form patterns and produce material and social effects' (Ahlborg 2017, p. 128). Influenced by feminist scholarship and working outside of the MLP, Ahlborg argues for understanding power as an emergent, productive and contingent phenomenon. One of the benefits of employing a productive interpretation of power, she suggests, is that it 'leaves more room for *contradictory and ambiguous outcomes*' (Ahlborg 2017, p. 127, original emphasis) than a focus on power 'to'. Similarly to the knowledge-politics perspective outlined in Chapter 1, Ahlborg argues that the stability of a socio-technical system cannot be explained simply in terms of domination by particular actors, for power 'over' is also expressed through the ways in which systems are configured.

Conclusions

Politics and power are critical considerations in the governance of urban transitions and various authors have emphasised the need to devote considerably more attention to these topics in order to advance transition scholarship and policy (e.g. Scoones *et al.* 2015; Patterson *et al.* 2017). In this chapter, I have examined the meaning of these concepts and their interconnections with knowledge production and consumption. Politics has been purposefully defined in a broad manner, using the Laswellian language of 'who gets what' to emphasise the breadth of phenomena that can be considered political. In putting forward an argument that knowledge matters politically, I have examined the gross social inequalities perpetuated through the development of the New York High Line Park, and linked these outcomes to the way in which the project was rendered, and re-rendered, knowable. Other contributors to this book greatly extend the analysis of the micro-politics of urban knowledge production and consumption in additional situated cases.

I have argued that an analysis of knowledge politics must necessarily centrally address questions of power. The multiplicity of ways in which power is understood can act as a barrier to such a goal and I have therefore attempted to demystify some of the main debates concerning power. The categorisation I have employed as an analytical heuristic to structure the discussion of power, needless to say, is somewhat simplistic. There are considerable overlaps between the categories used and they conflate a much greater plurality of perspectives than I can hope to portray in this chapter. Nevertheless, it is worth noting that in her recent contributions to the transition literature, Partzsch (2017a; 2017b) emphasises the value of using a multidimensional lens in addressing questions of power for it can help to open our eyes to the multiplicity of ways in which power is expressed and contested.

By examining that part of the transition literature that has addressed power explicitly, it is evident that there has been a strong focus on how human agency (constitutive of power 'to') can form a counterweight to domination by what is variously referred to as incumbent regimes, regime actors or the elite. There is a deep-rooted belief in the literature in the capacity of individuals and society to orchestrate change; indeed, it has been suggested that human agency is the characteristic that truly defines a transition (Chaffin *et al.* 2016). An assertion that intentionality is central to a sustainability transition is certainly valid, but it has been noted that foregrounding the capacity of actors to instigate change may risk obscuring other important components of societal power dynamics (Geels 2014). Arguments have also been put forward in the transitions literature for greater attention to be given to social movements and agonistic politics, to the political work involved in reproducing the status quo, and to the way in which relationships of domination/subordination are conceptualised. It is evident that a productive interpretation of power, which is central to our conception of knowledge politics, has received relatively limited attention to date, despite its popularity in geography and other disciplines.

Notes

1 The terms transition and transformation, and their synonyms, are used interchangeably in this chapter, although I acknowledge that for some individuals the two terms have distinct meanings. It has increasingly been argued, for example, that the term transformation more aptly reflects the scale of changes required (Patterson *et al.* 2017), not least given the systemic focus of transitions studies.

2 Power 'to' is typically interpreted as being a non-relational construct, but not all scholars conform to this understanding. Barrett and Duvall, for example, define power 'to' as a function of social relations of constitution that define actors relationships (Barnett and Duvall 2005).

References

Ahlborg, H., 2017. Towards a conceptualization of power in energy transitions. *Environmental Innovation and Societal Transitions*, 25, 122–141. Available at: https://doi.org/10.1016/j.eist.2017.01.004.

Allen, A., 1999. *The power of feminist theory: Domination, resistance, solidarity*. Boulder, CO: Westview Press.

Allen, J., 2003. *Lost geographies of power*. Oxford: Blackwell.

Arendt, H., 1970. *On violence*. London: Penguin.

Avelino, F., 2017. Power in sustainability transitions: analysing power and (dis)empowerment in transformative change towards sustainability. *Environmental Policy and Governance*. Available at: https://doi.org/10.1002/eet.1777.

Avelino, F. and Rotmans, J., 2011. A dynamic conceptualization of power for sustainability research. *Journal of Cleaner Production*, 19 (8), 796–804. Available at: http://dx.doi.org/10.1016/j.jclepro.2010.11.012.

Avelino, F. and Wittmayer, J.M., 2016. Shifting power relations in sustainability transitions: a multi-actor perspective. *Journal of Environmental Policy & Planning*, 18 (5), 628–649.

Avelino, F., Grin, J., Pel, B. and Jhagroe, S., 2016. The politics of sustainability transitions. *Journal of Environmental Policy & Planning*, 18 (5), 557–567. Available at: www.tandfonline.com/doi/full/10.1080/1523908X.2016.1216782.

Barnett, M. and Duvall, R., 2005. Power in global governance. In M. Barnett and R. Duvall, eds. *Power in global governance*. Cambridge: Cambridge University Press, pp. 1–32.

Bulkeley, H., Castán Broto, V. and Edwards, G., 2015. *An urban politics of climate change: Experimentation and the governing of socio-technical transitions*. London: Routledge.

Cataldi, M., Kelley, M., Kuzmich, H., Maier-Rothe, J. and Tang, J., 2011. Residues of a dream world: the High Line, 2011. *Theory, Culture & Society*, 28 (7–8), 358–389.

Chaffin, B.C., Garmestani, A.S., Gunderson, L.H., Benson, M.H., Angeler, D.G., Arnold, C.A., Cosens, B., Craig, R.K., Ruhl, J.B. and Allen, C.R., 2016. Transformative environmental governance. *Annual Review of Environment and Resources*, 41 (1), 399–423. Available at: www.annualreviews.org/doi/10.1146/annurev-environ-110615-085817.

Connelly, S. and Richardson, T., 2004. Exclusion: the necessary difference between ideal and practical consensus. *Journal of Environmental Planning and Management*, 47 (1), 3–17.

David, J. and Hammond, R., 2011. *The High Line: The inside story of New York City's park in the sky*. New York: Farrar, Straus and Giroux.

Foucault, M., 2002. *Power: Essential works of Foucault 1954–1984, volume 3*, J.D. Faubion, ed. London: Penguin.

Geels, F.W., 2002. Technological transitions as evolutionary reconfiguration processes: a multi-level perspective and a case-study. *Research Policy*, 31 (8–9), 1257–1274. Available at: http://linkinghub.elsevier.com/retrieve/pii/S0048733302000628.

Geels, F.W., 2004. From sectoral systems of innovation to socio-technical systems: Insights about dynamics and change from sociology and institutional theory. *Research Policy*, 33 (6–7), 897–920.

Geels, F.W., 2014. Regime resistance against low-carbon transitions: introducing politics and power into the multi-level perspective. *Theory, Culture & Society*, 31 (5), 21–40. Available at: http://journals.sagepub.com/doi/10.1177/0263276414531627.

Geels, F.W, Sovacool, B.K., Schwanen, T., Steven, T. and Steven, S., 2017. The socio-technical dynamics of low-carbon transitions. *Joule*, 1 (3), 463–479. Available at: https://doi.org/10.1016/j.joule.2017.09.018.

Green, J., 2013. High line founders win prize for their 'impossible' park. Available at: https://dirt.asla.org/2013/10/03/high-line-founders-win-vincent-scully-prize-for-their-impossible-park/ [accessed 10 October 2017].

Haan, F.J. De and Rotmans, J., 2018. Technological forecasting & social change: a proposed theoretical framework for actors in transformative change. *Technological Forecasting & Social Change*, 128 (December 2016), 275–286. Available at: https://doi.org/10.1016/j.techfore.2017.12.017.

Habermas, J., 1984. *The theory of communicative action. Volume one: Reason and the rationalization of society*, T. McCarthy, trans. Boston, MA: Beacon Press.

Habermas, J., 1987. *Theory of communicative action. Volume two: Lifeworld and system*, T. McCarthy, trans. Boston, MA: Beacon Press.

Haugaard, M., 2002. *Power: A reader*. Manchester: Manchester University Press.

Haugaard, M., 2003. Reflections on seven ways of creating power. *European Journal of Social Theory*, 6 (1), 87–113.

Haugaard, M., 2006. Power and hegemony in social theory. In M. Haugaard and H.H. Lentner, eds. *Hegemony and power: Consensus and coercion in contemporary politics*. Oxford: Lexington, pp. 45–66.

Hess, D.J., 2016. The politics of niche-regime conflicts: distributed solar energy in the United States. *Environmental Innovation and Societal Transitions*, 19, 42–50. Available at: http://dx.doi.org/10.1016/j.eist.2015.09.002.

Hoffman, J., 2013. Theorizing power in transition studies: the role of creativity and novel practices in structural change. *Policy Sciences*, 46, 257–275.

Hoffman, J. and Loeber, A., 2016. Exploring the micro-politics in transitions from a practice perspective: the case of greenhouse innovation in the Netherlands. *Journal of*

Environmental Policy & Planning, 18 (5), 692–711. Available at: www.tandfonline.com/doi/full/10.1080/1523908X.2015.1113514.
Jasanoff, S., 2004. *Design on nature: Science and democracy in Europe and the United States*. Princeton, NJ: Princeton University Press.
Jasanoff, S., 2015. Future imperfect: science, technology, and the imaginations of modernity. In *Dreamscapes of Modernity: Sociotechnical Imaginaries and the Fabrication of Power*. Chicago, IL: University of Chicago Press, pp. 1–33.
Jordan, A. and O'Riordan, T., 2001. Environmental politics and policy processes. In T. O'Riordan, ed. *Environmental science for environmental management*. Harlow: Prentice-Hall, pp. 63–92.
Kimmelman, M., 2014. The High Line opens its third and final phase. *New York Times*. Available at: www.nytimes.com/2014/09/20/arts/design/the-high-line-opens-its-third-and-final-phase.html [accessed 13 October 2017].
Lang, S. and Rothenberg, J., 2017. Neoliberal urbanism, public space, and the greening of the growth machine: New York City's High Line park. *Environment and Planning A*, 49 (8), 1743–1761.
Laswell, H., 1935. *Politics: Who gets what, when, how*. London: McGraw-Hill.
Latour, B., 1983. Give me a laboratory and I will raise the world. In K. Knorr-Cetina and M. Mulkay, eds. *Science observed: Perspectives on the social study of science*. London: Sage, pp. 141–170.
Leach, M., 2015. What is green? Transformation imperatives and knowledge politics. In I. Scoones, M. Leach and P. Newell, eds. *The politics of green transformations*. Abingdon: Routledge, pp. 25–38.
Lukes, S., 1974. *Power: A radical view*. London: Macmillan.
Mann, M., 1986. *The sources of social power: Volume 1 – a history of power from the beginning to AD 1760*. New York: CUP.
McGeehan, P., 2011. The High Line isn't just a sight to see: it's also an economic dynamo. *New York Times*, p. A18. Available at: www.nytimes.com/2011/06/06/nyregion/with-next-phase-ready-area-around-high-line-is-flourishing.html [accessed 10 October 2017].
Millington, N., 2015. From urban scar to 'park in the sky': terrain vague, urban design, and the remaking of New York City's High Line Park. *Environment and Planning A*, 47 (11), 2324–2338.
Mills, C.W., 1956. *The power elite*. Oxford: Oxford University Press.
Navarro, M., 2015. In Chelsea, a great wealth divide. *New York Times*, p. MB1. Available at: www.nytimes.com/2015/10/25/nyregion/in-chelsea-a-great-wealth-divide.html [accessed 13 October 2017].
O'Brien, K., 2012. Global environmental change II. *Progress in Human Geography*, 36 (5), 667–676. Available at: http://journals.sagepub.com/doi/10.1177/0309132511425767.
Ottinger, G., 2017. Making sense of citizen science: stories as a hermeneutic resource. *Energy Research & Social Science*, 31, 41–49. Available at: http://dx.doi.org/10.1016/j.erss.2017.06.014.
Pansardi, P., 2012. Power to and power over: two distinct concepts of power? *Journal of Political Power*, 5 (1), 73–89.
Parsons, T., 1967. *Sociological theory and modern society*. New York: Free Press.
Partzsch, L., 2017a. Biofuel research: perceptions of power and transition. *Energy, Sustainability and Society*, 7 (1), 1–10.
Partzsch, L., 2017b. 'Power with' and 'power to' in environmental politics and the transition to sustainability. *Environmental Politics*, 26 (2), 193–211. Available at: http://dx.doi.org/10.1080/09644016.2016.1256961.

Patterson, J., Schulz, K., Vervoort, J., van der Hel, S., Widerberg, O., Adler, C., Hurlbert, M., Anderton, M., Sethi, M. and Barau, A., 2017. Exploring the governance and politics of transformations towards sustainability. *Environmental Innovation and Societal Transitions*, 24, 1–16. Available at: http://dx.doi.org/10.1016/j.eist.2016.09.001.

Raven, R., Kern, F., Verhees, B. and Smith, A., 2016. Niche construction and empowerment through socio-political work. A meta-analysis of six low-carbon technology cases. *Environmental Innovation and Societal Transitions*, 18, 164–180. Available at: http://x.doi.org/10.1016/j.eist.2015.02.002.

Raven, R., Kern, F., Smith, A., Jacobsson, S. and Verhees, B., 2016. The politics of innovation spaces for low-carbon energy: introduction to the special issue. *Environmental Innovation and Societal Transitions*, 18, 101–110.

Reichl, A.J., 2016. The High Line and the ideal of democratic public space. *Urban Geography*, 37 (6), 904–925. Available at: http://dx.doi.org/10.1080/02723638.2016.1152843.

Scoones, I., Newell, P. and Leach, M., 2015. The politics of green transformations. In I. Scoones, M. Leach and P. Newell, eds. *The politics of green transformations*. Abingdon: Routledge, pp. 1–24.

Shove, E., 2010. Social theory and climate change. *Theory, Culture & Society*, 27 (2–3), 277–288. Available at: http://journals.sagepub.com/doi/10.1177/0263276410361498.

Stirling, A., 2015. Emancipating transformations: from controlling 'the transition' to culturing plural radical progress. In I. Scoones, M. Leach and P. Newell, eds. *The politics of green transformations*. Abingdon: Earthscan, pp. 54–67.

Temper, L., Walter, M., Rodriguez, I., Kothari, A. and Turhan, E., 2018. A perspective on radical transformations to sustainability: resistances, movements and alternatives. *Sustainability Science*, 13 (3), 1–18. Available at: http://dx.doi.org/10.1007/s11625-018-0543-8.

The Crown Estate, 2018. The Crown Estate: Our business. www.thecrownestate.co.uk/en-gb/our-business.

Weber, M., 1978. *Economy and society: An outline of interpretive sociology*, G. Roth and C. Wittich, eds. London: University of California Press.

3 Smart urbanism in Barcelona

A knowledge-politics perspective

Evelien de Hoop,[1] *Adrian Smith,*[2] *Wouter Boon,*[3] *Rachel Macrorie,*[4] *Simon Marvin*[5] *and Rob Raven*[6]

> Everything flows through the data centre in the middle of the town. So, if I own that data centre, I've got everything. Literally every part of the town comes right through there … Sold to me by Hitachi … I am the State. I got it all … It's kind of a packaged deal. Not just energy, or transportation, or home, or health, or whatever. They're actually ready to do what looks pretty much like a city council campaign. You could actually run for office like this … What are the political implications when you've actually sold off control over your streets; when you've offshored it to Silicon Valley or wherever it goes? Have you thought through the political implications of that? It may look smart, but how smart is that move? What happens if the App is on strike, or removes itself? How do you get the App back? Can you buy it; can you legislate it; can you code it yourself? You're over a barrel sister! I'm a smart city but my brain is run in California. How is that supposed to work out? Look behind the beautiful façade!
>
> (Bruce Sterling, cyberpunk, at the 10th International FabLab Convention in Barcelona)[7]

Introduction

Bruce Sterling's quote vividly captures some of the knowledge politics involved in smart urbanism. Or at least, the concerns critics raise against the Smart-City vision: the way urban processes are reduced to codified, inter-operable (and tradeable) information; whose processing through calculation and inference produces ostensibly authoritative knowledge about complex cities; and whose proprietary characteristics cedes power to 'smart-city' service providers (Greenfield 2013). At the same time, however, technological advances in sensors, data handling, internet platforms, ubiquitous computation and evermore imaginative visualisation permits wider access to information about cities and opens up the possibility for unprecedented citizen involvement in urban processes.

Barcelona was an appropriate place for Bruce Sterling to make his remarks. He did so on a conference platform that later that day saw then-Mayor Xavier Trias commit Barcelona to becoming a smart, self-sufficient city within 40 years (see later). Under Mayor Trias, the city government was working hard to promote Barcelona as a world-leading smart city, and with considerable success (Continente *et al.* 2016). A wide variety of smart-city installations had been implemented; international smart-city service providers were locating their business operations into a test-bed district; and an ecosystem of smaller developers and start-ups were innovating smart Apps and other tools.[8] Work was underway to integrate this patchwork of smart-city elements into what was called an 'Operating System' (OS) for the city: the 'essential hardware, software and data components that quietly sit in the background directing urban flows, providing shared languages towards interoperability across multiple infrastructures' (Marvin and Luque-Ayala 2017, p. 1).

Then, in 2015, city elections provided a moment for recent shifts in the urban political landscape. The new Mayor, Ada Colau, took office with a vision rooted in citizen mobilisations, commons and collaborative approaches to urban experimentation, and prototypes for direct democracy. Her new party, Barcelona en Comú [Catalan for Barcelona in Common] emerged from a kaleidoscope of innovative practices carried by the burst of activism by the 15-M movement over the period 2011–2015, and which emerged in response to the political and economic crisis triggered by the 2008 financial crash. Barcelona en Comú won 11 out of 41 seats, and operates within a minority government. Under this coalition, the smart city was no longer a priority. Technological sovereignty has taken its place within a broader agenda for more democratic urban developments. In the terms of the Introduction to this book, Barcelona has undergone a shift in political rationality: 'particular ways of thinking about, calculating and responding to [urban] problems' (Chapter 1, p. 9). Propelled by changing realities in the city and the emergence of new protagonists brought forwards by the contradictions in those realities, contending political rationalities have had consequences for the design and implementation of digital knowledge-producing techniques.

In this chapter, we look to Barcelona as a case study in the knowledge politics of smart urbanism. Barcelona helps us explore an issue central to this book, which is how 'the political processes through which knowledge-configuring practices and strategies for representing the urban fabric are assembled and institutionalized' (Chapter 1, p. 2). What is fascinating in Barcelona, and we suspect other cities, is how digital projects can be disrupted, reconceived and reclaimed – or complemented and replaced by new digital projects – through urban politics that interact across elite and grassroots settings, and in ways that suggest more plural and hopeful possibilities.

Smart urbanism

The smart city is not a new response to urban development, but the latest in a stream of initiatives over the last half-century to incorporate information and communication technologies into processes for knowing and governing cities (Marvin and Luque-Ayala 2017). Often, the promise of such radical technological changes is to rationalise urban socio-technical transformations, and the implication is that it risks entrenching a 'post-political', technocratic approach to the city (Swyngedouw 2009). Critical urban scholars see in smart-city visions and experiments the interests of a rather narrow neo-liberal agenda focused upon technological fixes controlled by coalitions of corporate actors and city elites (Greenfield 2013; Kitchen 2014). As such, cities cede authoritative urban knowledge production, which should be subject to democratic scrutiny, to commercially interested technology providers. In this line of argument, ownership and control of knowledge-producing and urban-governance technologies are key political issues (Feenberg 1999). Yet, advocates argue that ceding authoritative urban knowledge production to private parties is a reasonable political arrangement if those commercially interested smart-city service providers produce knowledge that improves urban performance for their public authority clients, and if citizens benefit from improved services and more efficient use of their tax contributions. In this controversy between advocates and critics of the smart city, knowledge politics emerges with regard to the ownership and control of the knowledge-production apparatus, its credibility and its performance.

Advocates and critics alike tend to see smart-city technologies as largely impacting upon cities, like an external force of change, for good or ill. Sometimes there is recognition that smart technologies are shaped by global social and technological processes, such as neo-liberal ideology, public research into computation and Big Data, the rise of Silicon Valley, or normalising digital cultures. These social forces are usually deemed so pervasive or powerful that, from the perspective of an individual city, they appear as all-determining (Jordan 2008). However, there remains relatively little consideration for the way in which digital technology developments may take shape in specific socio-material urban settings. Urban actors and historical social and material heritage exercising agency – which we will refer to as urban agencies – over digitally enabled ways of conceiving, knowing about and intervening in urban development tend to be overlooked or downplayed.

Indeed, grassroots initiatives have been developing alongside off-the-shelf packages for smart-city services, for instance in the area of sensors and environmental information; a development that potentially opens uses of ICT for citizen-led urban governance (Gabrys 2015; Tironi and Sanchez Criado 2015). Arguably, more democratic control of smart-city technologies, including the utilisation of free and open software protocols, permits pluralistic values and assumptions to continually enter governance debates and deliberations (Kurban *et al.* 2017). Furthermore, whilst powerful market and ideological dynamics shape ICT developments into forms that centralise urban knowledge convenient to corporate and urban elites, the implementation of off-the-shelf packages requires adaptation to a particular city's characteristics (Kahn and Kellner 2007). Global social forces are therefore always confronted with local dynamics, which therefore need to be taken into account to understand smart urbanism. Municipalities are also developing expertise in-house, and working with grassroots groups, local businesses and larger providers to develop bespoke services (Hajer and Dassen 2014; Marvin and Luque-Ayala 2015). Recognising urban agencies in smart-city developments highlights that a much more hybrid, less monolithic smart urbanism is emerging, each nevertheless involving an inescapable knowledge politics.

Overlooking urban agencies in favour of focusing on global social and technological processes shaping smart technologies also obfuscates the agency of smart technologies' materialities themselves – rather than their ownership and control – in the production of knowledge for urban governance. As this book's introductory chapter argues: the production of knowledge is rooted not only in discourse but also in material epistemic technologies: material artefacts enacting and articulating certain ways of knowing the city. The realisation of smart city's promises relies upon the production, analysis and integration of big and varied datasets, produced by a plethora of sensors (including static devices embedded in city infrastructures and mobile devices carried by citizens), and run by computers and servers whose calculations produce representations (ideally in real time) about different aspects of city performance. Knowledge is produced through networked platforms: city-dash boards, Apps, control centres and other devices that enable people to know what is going on in the city, who can intervene to adapt behaviour accordingly. These devices may afford different kinds of human participation: the devices enact people and other things in certain ways (Tironi and Sánchez Criado 2015). Some smart-enabled interventions are highly automated, such as the optimisation of traffic systems, or the triggering of actuators linked to, say, irrigation of parks. In other cases, human users are required to respond to signals, nudges or incentives issued by the smart-city system, effectively acting automatically if human response involves little reflection. And then there are smart services that invite more active deliberation and decision by 'smart citizens', particularly when platforms are open to adaptation, but nevertheless within an invited arena with designed parameters for participation (Vanolo 2016; Kurban *et al.* 2017).

Knowledge emerging from smart technologies is not only political because this knowledge may play an active role in urban governance decision-making, but also because the process of knowledge creation inevitably involves making choices, an insight emerging from science and technology studies (STS; see Chapter 1). Whilst there are distinct schools of thought, at its heart STS analyses how any 'techno-scientific' project intended to better understand and govern complex worlds, necessarily carries normative assumptions and values in its design and implementation (Jasanoff 2004). The knowledge produced by a smart-city project is constructed through the interaction of design choices, user practices, the platforms themselves and the agency of the material characteristics of the issue about which knowledge is being produced (Latour 2005). In other words, knowledge production in smart platforms is a political process, involving multi-scalar negotiation, agreement and dispute about the implementation of smart-city technologies. This effectively determines what information is gathered, how knowledge is produced (as well as inevitable gaps and uncertainties) and interpreted, and then how these insights are represented as reliable knowledge for acting in the city (Matthewman 2011).

From this short review, the following three aspects of smart urbanism's knowledge politics can be distilled: (i) the critical role of urban agencies – or understanding smart urbanism as a situated, socio-material practice; (ii) the agency of smart-city technologies' materiality as well as the ownership and control of these technologies, and: (iii) the political rationalities, values and assumptions embodied in their design and use. We will use these insights as a lens on Barcelona as a smart city in order to better understand the interrelation between smart knowledge-configuring practices and shifting political rationalities.

Barcelona: moving beyond the smart city?

Barcelona is often presented internationally as a pioneer in modern urbanism. The Cerdà plan extended (and liberated) the city beyond its walls in the 1860s in a way that became symbolic for modernisation (Aibar and Bijker 1997). Soon after the walls fell, the city fortress was demolished in order to create space for a Universal Exposition in 1888 (and what became the Ciutadella Park). The Exposition showcased urban redevelopments and a modern city to over two million visitors. A second World's Fair was held on Montjuïc in 1929. Developments for this exposition, conceived as a further motor for city development, similarly produced important monuments to modernisation. This 'model' evolved fitfully over the years – an international projection of the city based in modernising district developments – and found spectacular expression with the Olympic Games in 1992, and a less successful attempt with the Universal Forum of Cultures in 2004 (Degen and García 2012).

Implementation of the 'Barcelona model' (Ribera Fumaz 2017) always involved political controversy and social and economic struggle (Aibar and Bijker 1997; McDonogh 2011): between social classes, over space, in the priorities and directions for development, concerning rights to the city, contradictions between attracting inward investment and attending to neighbourhood needs and, pertinent here, the selective use of technology and knowledge about the city. Confronting the models of Barcelona's urban elite is a grassroots history of working-class struggle, neighbourhood and nationalist politics, and social movements, all of which have understood their city quite differently. These contentious histories provide an important context for appreciating the arrival of the Smart City in Barcelona, and an aspiration for technological sovereignty.

The smart-city vision

The positioning of Barcelona amongst the vanguard of Smart Cities was attempted most explicitly and vigorously under the leadership of Mayor Trias (2011–2015) (Continente *et al.* 2016). Trias's Deputy Mayor for Urban Habitat, Antoni Vives, and the department's general director Vicente Gaullart set about articulating and implementing the vision. Prior to entering public office, Vives and Gaullart had been co-founders in 2001 of the Institute of Advanced Architecture of Catalonia (IAAC), dedicated to propel research and education about bringing new technologies into urban spaces. Whilst drawing upon visionary ideas cultivated at IAAC, these smart-city leaders capitalised on projects and skills already present in Barcelona. In doing so, they articulated and repackaged initiatives already underway in the city, attempting to reorientate the city to their vision.

In terms of operations and strategies already present in the area, the city had in 1990 created the Municipal Institute for Information Technology (IMI – Institut Municipal d'Informàtica) as an independent municipal body providing IT services. Fibre-optic cable began weaving through the city from 1994, and the introduction of networked computing services promoted thereafter. In 2000, the previous city council had announced an ambitious plan to develop a Knowledge District in the Poblenou area. Informed by a second phase of regeneration planning after the Olympics, this relatively run-down and poorly served district was earmarked for residential, educational and commercial developments. The objective was to turn it into an attractive site for international investment in a knowledge economy that would attract a high-skill workforce to the city, including the ICT sector (Leon 2008). The district was called 22@ and a municipal company, also called 22@, was created to market and manage the real estate and infrastructure projects involved. International property developers and investment firms began investing in the buildings rising above the old factories, mechanics and artists workshops, and housing.[9] The Torre Llacuna tower building, iconic for 22@, was completed in 2003. The nearby Media-ICT building won the World Building of the Year award in 2011 (March and Ribera-Fumaz 2014, p. 5; Continente *et al.* 2014). Local universities opened campuses in the district. Barcelona Activa, the economic development agency of the city council, turned its attention to high-tech start-ups and entrepreneurs, and opened offices and provided incubation space in the Media-ICT building.

Under Mayor Trias, Barcelona committed to Open Data policy in 2011, and began providing citizens and businesses with access to some of the information gathered by the council about the city. The city was selected as the Mobile World Capital by GSMA in 2011, and hence hosted future Mobile World congress events as well as the Smart City Expo in the same year. These events brought business executives and urban policymakers to the city. Moves were made to attract international smart-city technology firms to locate their development operations in 22@. Strategic agreements were announced, for example, with Cisco, IBM, Schneider-Telvent, Telefónica, GDF Suez and others. Indeed, as the smart city vision developed, so 22@ became marketed as a campus for smart city developments (March and Ribera-Fumaz 2014; Charnock *et al.* 2014).

Smart systems were in place, or being established, to monitor and control traffic, energy, water use in parks, waste collection services, social services, care for the elderly, public transportation, and so forth (Bakici *et al.* 2013; Kuyper 2016). IMI and others were developing software systems for managing the accumulation of thousands of diverse of sensors (e.g. the Sentilo open-software platform). Leaders invited, and incentivised, technology developers to embed themselves in the city and conduct smart urban experiments. Local entrepreneurs and start-ups were encouraged to make use of the data and platforms that were emerging, and to participate in smart city projects (Capdevila and Zarlenga 2015). The city announced ambitions to create a city Operating System that would interconnect information from across the multiplying sensor networks and data-gathering platforms in different city administration departments, and hence boost the ability of city authorities to observe and manage their intelligent city in real time.

Intriguingly, Vives announced, and began implementing, a plan to open ten public digital fabrication workshops in each of the ten districts of Barcelona as part of the smart-city vision. Inspired by a 'FabLab' model originating at MIT (and which IAAC had pioneered on its premises since 2006), these 'Ateneus de Fabricació Digital' were equipped with a suite of digital design and fabrication technologies that would allow citizens to train and experiment in the prototyping of physical objects, and to share designs, knowledge and collaborate digitally with similar workshops globally (Smith 2015). A network of neighbourhood Ateneus was envisaged as part of the public infrastructure for the twenty-first-century smart city. These production facilities were envisaged as enabling Barcelona to become more materially self-sufficient whilst competing in a more sustainable world: sustainable design knowledge will be traded globally, but physical goods were made (and remanufactured) in local circular economies (Diez 2012). At the tenth FabLab convention held in Barcelona in 2014, Mayor Trias committed his city to making over half the goods it consumed locally within 40 years (the smart, self-sufficient city; see Guallart 2014).

The ambition with the extensive portfolio of smart initiatives was to simultaneously attract and finance industrial development, such that Barcelona would become a global hub for the development of (exportable) smart-city services, whilst modernising the city itself for a smart citizenship – a citizenship comfortable using technology applications, responding to the data that they willingly contribute towards smart-city services, but not necessarily demanding other rights and responsibilities. By July 2015, *Fortune* magazine was writing how Barcelona was 'the most wired city in the world' (Walt 2015). The city regularly featured at the top of global rankings for smart cities and innovative cities.

And yet, throughout these developments, citizens of Barcelona were positioned as relatively passive beneficiaries of smart-city developments: as users of the services provided, beneficiaries of the visionary developments, workers in the new sectors; but, apart from the tech entrepreneurs, citizens actively involved in initiating and shaping grassroots 'smart' developments were overlooked. Meanwhile, myriad contracts were struck with businesses to develop the technologies and services, and demonstration funds from the European Commission and elsewhere were used to help embed smart devices in urban development. As *Fortune* magazine observed, '[d]iscreet and largely unannounced, the changes in Barcelona have slipped by even observant residents … Yet the stealthy transformation is profound and potentially so sweeping that no one is sure where it will lead' (Walt 2015). In fact, announcements were made, and marketed very professionally, but appeared to be addressed as much, if not more, to international audiences relevant to Barcelona's global position, than to local users of the new services.

Grassroots digital urbanism

Whilst the smart city was being embedded in bus-stops, lamp posts, refuse bins, mobility systems and machine-readable databases, and whose sensors and actuators were being interconnected through a developing Operating System, another smart Barcelona was becoming increasingly vocal and active. The economic crisis signalled by the financial collapse of 2008, and whose political consequences overwhelmed Trias' predecessor, was now troubling his administration in turn.

Construction, housebuilding and urban developments had fuelled much of the economic growth in Spain from the 1990s, until the bubble burst in 2008 (García 2010). The exposure of Spain to the global financial crisis was exacerbated by an associated implosion of its property market. Many families were left unable to meet mortgage repayments and, by 2012, one household was being evicted every 15 minutes; even though there was little social support for the homeless, and with 3.4 million homes vacant (Romanos 2013). The Platform of Mortgage Victims (PAH, La Plataforma de Afectados por la Hipoteca, formed in 2009) articulated moral outrage at this situation and mobilised it into effective practical action. In doing so, it became a citizen mobilisation significant beyond housing, in Barcelona and hundreds of other cities, and that joined a confluence of actions in energy, mobility, culture, urbanism and communications arising from 15-M in 2011. In the streets, and through varied actions, a broader vision for democracy and urban governance began to be articulated that was quite different to Trias' smart urbanism, and political programmes to put the broader vision into effect were developed (Gonick 2016; Gutiérrez 2017).

PAH built a power base that enabled it to renegotiate housing contracts and debt repayments on behalf of individual members. PAH did this by reframing evictions as a systemic and institutional issue rather than an individual failing, and coordinated more conventional campaigning activities for housing reforms alongside its practical help and solidarity for individual households (De Weerdt and Garcia 2016). Digital tools helped in the organisation and communication of this activity at scale: making decisions, operating transparently, sharing information, eviction alerts, and experience about legal and banking matters, and networking between neighbourhood associations, especially when the campaign moved rapidly and began to operate nationally.

Grassroots groups were developing initiatives in what was becoming known widely as the solidarity economy: loosely affiliated systems of cooperative production and consumption initiatives in areas as diverse as food, mobility, energy, culture, housing and politics. A variety of digital tools and platforms were being developed by activist coders and grassroots groups for assisting and coordinating solidarity and cooperative economic activity. Digital social innovations helped advance offline activities in citizen urbanism. Included in movement thinking was a rejection of neo-liberal models of globally competitive, marketised cities and, in the context of urbanism, new ways of thinking and doing urban practices dedicated to prototyping open neighbourhood developments, and through this the production of common goods and services (Estalella *et al.* 2013).

Experimental platforms were being built, typically involving an ethic of openness and collaboration inspired by free software and free culture movements, for connecting local producers to consumers, promoting ethical and solidarity exchanges, and anticipating different kinds of digitally enabled currency, peer-production and collaborative consumption. Digital citizen-deliberation techniques were emerging from the multitude of groups, networks, platforms and associations throughout Spain (2011–2013) (Monterde *et al.* 2015; Gutiérrez 2017). Activists in Barcelona worked in collaboration with activists in Madrid, A Coruña, Sevilla and other cities in Spain. At the same time as developing tools, there were debates and discussions about the conceptual frameworks for organising this activity, and for doing so strategically. Threading together the kaleidoscope of initiatives were commitments to direct democracy, transparency and citizen-centred forms of governance, and borne of indignation towards an incumbent system deemed (in the context of revelations of political and business corruption emerging after the economic crisis) to be inept, corrupt, closed and self-serving (Gutiérrez 2017).

The new visions for direct democratic urban practice were inspired by affinities between concepts coming from autonomist and feminist movements, free-culture movements and, significantly, the free-software movement. A hacker ethic towards decentralised forms of online coordination, collaboration and peer production, and commitment to transparent and democratic rights in technology, informed the conceptual thinking. If the smart-city vision subsumed (and shaped) digital affordances within a logic of markets, service provision and neo-liberal ideas about urban efficiency, then here, activists were shaping digital affordances and applying them within a quite different framework. A generation accustomed to accessible digital technology speeding up their ability to communicate and organise, took this practice in a seemingly more directly democratic direction. Social media tools were adapted into 'technopolitical media' for: collecting ideas; discussing them, opposing them and modifying them; voting and taking decisions; and communicating this through content creation using new media (Kurban *et al.* 2017). These tools proved promising and helpful, and took the concept of digital cities in quite different directions compared to the closed development of dashboards and platforms by technology corporations, and in which data is centralised, released conditionally and wraps a veneer of transparency and controlled participation around existing urban institutions. Moving beyond these elite forms of e-government, technopolitical activists were working at canvassing citizen knowledge to build more direct decision-making and citizen-led urban governance, and connecting citizens as protagonists in new relations with representative democratic institutions (Kurban *et al.*

2017). These digital platforms relied upon voluntary contributions of knowledge and skill, and crowdsourced funds. The code for tools, developed originally to organise 15-M occupations, and for coordinating and debating developments in initiatives and movements thereafter, was shared through GitHub, such as the Consul free software. Early platforms like Propongo, and tools used by Partido X and Podemos, went on to underpin city platforms like Decide.es, and that are now being used by a variety of city administrations.

All this was happening at the same time as the international promotion of Barcelona as a smart city. Activists saw two different cities: the elite Barcelona using smart city as a brand in its neo-liberal competition for capital – a city rendered into an efficient and convivial location for mass tourism and the global knowledge economy; and the Barcelona of neighbourhood activism, struggling to build from below what they considered to be a more democratic urbanism capable of addressing issues and problems considered inherent to the neo-liberal model. Tellingly, when the city council moved to open one of its first Ateneus de Fabricació Digital in the disadvantaged neighbourhood of Ciutat Meridiana in 2013, for example, the building they chose was already being used by a community food bank. Neighbours occupied the building in protest: they needed food and local solidarity, not 3D printers and design platforms. A resolution was found in rehousing the food bank and committing the Ateneu to training young people into work. The event illustrated a clash between the citizens envisaged by the Council in the future smart city with the pressing realities confronting citizens today (Smith 2015).

The example of decision-making platforms illustrates how various technopolitical actors were helping connect traditional forms of urban activism to new scales and forms of agenda-setting, decision-making and mobilisation amongst citizens, and in so doing create new ways of knowing and acting in the city. Mistakes, lessons and knowledge production could proceed rapidly: about local activities, discussed, shared, reframed, aggregated with knowledge about similar initiatives elsewhere, and mobilised and organised for a more participatory, commons-based and democratic kind of urbanism. Platforms provided new ways of engaging and connecting with more traditional forms of neighbour mobilisation. Activists were practising an incipient urban governance operating beneath and around existing institutions. Ultimately, however, some activists argued that if democratic technopolitics – the subversive use of technologies combined with legal and political tools – was to realise its potential in less precarious and more influential ways, an engagement with city institutions was necessary (Kurban *et al.* 2017). New parties like Podemos and Barcelona en Comú were formed from amongst activist networks as a strategy to enter and transform urban institutions (Eizaguirre *et al.* 2017).

Technological sovereignty

Democratic technopolitics was instrumental to the development of the new political parties that emerged from the 15-M social movements, including

Podemos and Barcelona en Comú. Platforms were used to select candidates, publish financial information, set agendas, debate issues, communicate across meetings, decentralise campaigns to local groups, communicate with allied groups, networks and movements, and keep leaderships accountable to the horizontal networks that constituted their power base. Working in this way in 2015 city elections (Eizaguirre *et al.* 2017), Barcelona en Comú won enough votes to form a minority city government in partnership with other parties. Ada Colau, who had been a spokesperson for PAH, became Mayor.

Smart urbanism was not a priority compared to issues of housing, corruption, mass tourism, rights to the city and aspirations for more commons-based economic activity and the remunicipalisation of core city services. Nevertheless, digital tools for public deliberation were emblematic of aspirations to create a new way for city politics. Moreover, the underlying vision for approaching these and other issues, seeking commons-based developments through citizen participation, were informed by ideas in free culture and commons-based urbanism similar to those inspiring democratic technopolitics. The smart city of the previous administration came to be reconsidered through a technopolitical lens, and digital urbanism in the city reoriented to a policy of technological sovereignty. Another strand to the technological sovereignty vision was mounting concern about the concentration of data in the hands of a handful of poorly regulated technology-platform providers. Wikileaks revelations of state interest in this data similarly heightened concern about data and control over technology. Questions of data sovereignty and the capacity of citizen rights in basic information utilities were a topic moving beyond digital activism and into the public eye.

It took time to develop a digital strategy aligned to the democratic urban commons towards which the new city government aspired. Studies were commissioned and people appointed, including Francesca Bria as Chief Technology and Digital Innovation Officer, who had a background in European digital projects aligned to technopolitical aspirations. Some of these projects, such as the D-CENT citizen-deliberation platform, involved technopolitics activists based in Barcelona. The new Digital City Plan for 2017–2020 was published in October 2016. The leitmotif for the new plan was to move beyond the smart city and transition to technological sovereignty. Transparent digital technologies accountable to citizens were sought in the strategy, that would contribute to urbanism that is commons-based, circular and creative (Barcelona City Council 2016). Council policy is to develop a data commons, and to use this transparently through open-source digital practices. It involves a redirection and redevelopment of the Operating System projects inherited from the previous administration.

Emblematic projects have been established to implement the plan. Decidim. Barcelona has taken the technopolitics platforms for activist agenda-setting and decision-making and developed these into a platform for citizen participation in urban policy and development.[10] In keeping with the activist ethos, not only is the platform open source, but so too is the process for reflecting upon its operation and future development through a 'meta-lab' of open debate. This includes discussion about connecting online deliberation with face-to-face discussion and activity: bringing social weight to online information through the involvement of urban groups and the organisation of neighbourhood meetings. Decidim benefitted from earlier work in the D-CENT European Commission project for developing direct democracy platforms and which involved Francesca Bria, then at Nesta, and other European partners. A further European project, called DECODE, involves the city council digital team working with European partners in the development of blockchain techniques aimed at giving citizens greater control over their digital identity, with tools to help them set the way their data is shared for public good purposes (cf. corporate models of data extraction). Meanwhile, Decidim has been used to open up council processes to citizen participation, in the form of canvassing ideas and suggestions, coordinating comment and discussion, and voting on decisions, and which have input to planning documents, local initiatives, budgeting and other areas. It is currently being used in a process to rethink the development of the 22@ district and open it to citizen proposals. The aspiration is for new ways for people to participate in knowing and shaping the city: not as data points, but as co-designers.

Inserting these new design requirements into contracting arrangements with digital service providers has been part of the technological sovereignty policy. The reform of public purchasing policy to require open software and see data ethics encoded into techniques that give each citizen rights over their data is a key measure, although it is as of yet unclear what kinds of changes these requirements will induce. It is not always possible to renegotiate arrangements to the proprietary contracts struck by the previous administration, and so not all digital governance is retrospectively sovereign to council and citizens. Increasing diversity amongst vendors and providers, and not being reliant upon a limited number of packages, is another way the city authorities are trying to assert sovereignty. The promotion and opening up to digital social-innovation providers, for example, and taking policy cues from free-culture approaches, involves city authorities inserting values and assumptions that differ from those of the previous regime into digital infrastructures and services. Digital methods are being applied to the issues central to Barcelona en Comú's political vision, such as Big Data techniques for advancing socially inclusive housing and addressing illicit holiday lets, for example; or ensuring participation in the design of neighbourhood projects; or the commitment to transparency in public administration and politics.

Such diversity and more hybrid arrangements work to shift political relations in digitally enabled knowledge about the city. But the technology sovereignty vision is nevertheless tempered by the necessity of remaining competitive in a global digital economy. The strategy is careful to go beyond the smart city, while not rejecting smart urbanism. The plan speaks to diversify the digital economy. Agreements and provisions are made to continue promoting Barcelona as an attractive investment location for technology firms. Barcelona Activa remains an important hub for entrepreneurship, but now includes a team dedicated to supporting the solidarity economy, which helps groups pioneering digital social innovation in the city. The promotion of the city to capital continues with Smart City Expo and Mobile World Congress. There are policies for promoting skills and facilities for Industry 4.0, at the same time as support for a Makerdistrict and continued promotion of the Ateneus de Fabricació Digital. This raises interesting questions about the flexibility in which some knowledge-producing activities can easily move into the technological sovereignty rationale with little apparent change in practices. Programmes and initiatives to enable active citizen involvement in digital commons can equally equip them to work in the digital economy.

In all these changes in the political rationalities in relation to smart urbanism projects, it is important to remember that political leadership in the technology sovereignty strategy comes from a new political party borne of social movements, yet whose minority position in the council leaves it dependent upon agreements with other parties with differing 'political rationalities' (see Chapter 1). Moreover, the elected city council inherits city institutions and administrative bureaucracy and materialities (the Operating System, for example, and contracts already in place) shaped by historic layers of complex economic, social and political compromises not of their own making, and with which the digital strategy has to negotiate.

Not all urbanism is viewed through a digital lens, whether sovereign or smart. We write this chapter during the crisis in relations between Catalunya and the Spanish state, in which older forms of politics are dividing opinion and playing for advantage, and while Barcelona en Comù is a minority government that recently became even smaller after a recent fallout with one of its coalition partners. Highly charged contentions over sovereignty, democracy and governance are at stake, in which Barcelona's future rests.

Competing rationalities of smart city futures

The picture that emerges from our theoretical review and the example of Barcelona suggests more nuance is needed compared to the smart-city solutionism of advocates and the troubling hegemony seen by critics. Smart-city technologies do not land in cities as packages for installation or domination but are actively shaped by specific urban situations that have histories and geographies; in which smart-city initiatives negotiate complex institutional layers, and where digital initiatives are shaped and altered as much as digital imperatives shape those contexts.

In Barcelona, we see two major, contesting political rationalities for digital urban development, each emerging from global discourses as well as Barcelona's socio-material history and the specific life histories of key urban actors. Under Mayor Trias, a smart-city marketplace emerging from the corporate world was being promoted as an attempt to revitalise the competitiveness of the urban economy combined with programmes in which citizens were invited to train and experiment in the use of digital technologies. This rationality draws on

Barcelona's history as a showcase model for modern urbanism, which is tangible in Barcelona's present-day socio-material urban fabric, for example through its ongoing role as host of events such as the Smart City Expo, the material presence of the Cerdà plan in Barcelona's street layout, and the legacy of earlier international urban events. It is also inspired by the personal experience of Antoni Vives and Vicente Gaullart, who had been co-founders in 2001 of the IAAC, which was dedicated to advancing research and education on the use of new technologies in urban spaces. On the margins, and later under Mayor Colau, this position was challenged by another political rationality, which rejected neo-liberal models of globally competitive, marketised cities. Within the context of Barcelona's housing crisis and drawing on the city's history of social movements, this alternative was committed to direct democracy, transparency and citizen-centred forms of governance (Gutiérrez 2017; Eizaguirre *et al.* 2017). For its engagement with digital technologies, this movement drew on a hacker ethic, working with decentralised forms of online coordination and open-source technologies, committed to transparent and democratic rights in technology.

As part of these different political rationalities, different kinds of smart knowledge-configuring technologies were created. An important point to take from this is that knowledge politics figures around a multiplicity of urban transformations, not a single transition process. Under Trias, systems monitoring and controlling traffic, energy, water use in parks, waste collection services and more were put in place, which were to be combined in a City Operating System under uncertain and complex arrangements of ownership and control between city authorities and companies supplying these technologies (Bakici *et al.* 2013; Kuyper 2016). Such systems create knowledge based on the relatively passive participation of people and objects, whose behaviour is observed and measured using sensors and algorithms. These technologies were not introduced throughout the entire city, but in specific sectors, issues and/or districts amenable to prevailing investment and development opportunities. And, on the margins, there were resourceful spaces for activists and social entrepreneurs to develop alternative digital practices (including European Union programmes[11]), and which have also been infiltrating smart urbanism (Tironi and Sanchéz Criado 2015; cf. Marvin and Luque-Ayala 2017). These examples illustrate that urban agencies always need to be taken into account when trying to understand smart urbanism, also when the influence of neo-liberal ideology and international corporations is strong.

Urban agencies are even more tangible in the case of the smart knowledge-configuring technologies that emerged among grassroots movements and later when Barcelona en Comù took minority control of the council. This episode shows that political rationalities and smart knowledge-configuring technologies intersected and co-evolved, rather than one informing the other unidirectionally. Furthermore, the technologies that were created among grassroots movements at the time of the housing crisis are materially very different from the sensing and monitoring systems put in place under mayor Trias, and through their material difference facilitate different kinds of politics. Initially, grassroots movements appropriated open digital technologies for coordination and communication, based on principles of openness, accessibility, collaboration, decentralisation and sharing. These technologies helped to organise effective direct action and to develop alternative political visions. Digital tools were also being developed to assist and coordinate solidarity and cooperative economic activity, for example connecting local producers to consumers. Changes in control of the city council in Barcelona then opened up space for grassroots approaches to move beyond prototyping, beginning to institutionalise. The city's new digital strategy seeks technologies accountable to citizens, technologies that contribute to urbanism that is commons-based, circular and creative (Barcelona City Council 2016). Decidim.Barcelona was developed, facilitating communication with established institutions on policy issues through a combination of online deliberation with face-to-face discussion (Kurban *et al.* 2017). Commitments to values of openness, transparency and participation materialise through the design of the platform: it is open-source, and there is space for reflection on the platform's operation and future development. The creation of knowledge through Decidim.Barcelona and other platforms relies on the voluntary contributions of participants, who gain, rather than lose, agency through their participation. The politics of knowledge here spills beyond a powerful agent introducing synoptic technologies that consolidate and extend the framings of dominant interests in urban development: digital technologies are opened up by dynamic and contradictory initiatives in the city, between different urban actors and developments, and smart technology becomes a strategy for redefining and redistributing power relations in urban development. Indeed, the Barcelona case shows that a plurality of political rationalities sought different expectations, values and assumptions in the design of digital services, which had to be negotiated in the creation and use of technologies in the city.

Concluding remarks: avenues for future research

It is important to be reflexive towards the analysis presented here. Our approach to understanding the knowledge politics of smart urbanism looked at how contending ideas and experiences in urbanism, linked to shifting political control in the city, shaped and were shaped by the development of digital apparatuses for knowing and acting in the city. Our own knowledge-producing apparatus (our analytical framework and method) was productive for this purpose, but also overlooks important aspects of knowledge politics in smart urbanism. Our framing at the level of the city as a whole meant that the development consequences of fixing specific urban subsystems through digital means was outside our focus. To what degree, for example, do investments in specific smart systems entrain a degree of path-dependent developments that are difficult to dismantle or reorientate, should that be desired or required? Recoding and retooling is no straightforward matter once sociotechnical configurations stabilise and institutionalise. Indeed, technologies have a tendency to hardwire social relations

(Latour 1991), and to act as a source of inertia or even path-dependency over social relations seeking reforms (Kemp *et al.* 1998).

Furthermore, it is clear that further in-depth analysis of the enactment of smart technologies, such as sensor networks, or city databases or digital fabrication facilities, are required to assess the material consequences of core functional logics in digital approaches, to understand whether and how the computational logic of these smart technologies is susceptible to changes in urban rationale, and which knowledges are bracketed out or included in re-aligned digital codifications.[12] For example, knowledge about sustainability appears in reduced form in the smart-city and technology-sovereignty strategies in Barcelona. Notions of a self-sufficient city in the previous administration included an ecological modernist notion of how digital technologies will manage better urban metabolisms and close the loops of production and consumption through digital fabrication. Commitments to a circular economy in the technology-sovereignty strategy similarly see sustainability in terms of managing material flows and metabolisms. The material and social footprint of digital technologies themselves are glossed over. Any risks that digital infrastructures lock societies into accelerated cycles of upgrades and obsolescence, and whose material implications are troubling, are currently skirted around by digital policy. Knowledge that critiques ecological modernisation (York and Rosa 2003), and that problematises innovation as a solution to sustainable development, is not integral to the digital apparatus of knowledge production. Framings of urban sustainability that see socio-ecological relations as more than informational, and that propose less instrumental, more experiential and embodied ways of knowing and sensing, do not (yet) connect to digital framings and knowledge-producing apparatus. There may be affinities as well as tensions: for example, can we conceive of digital technological sufficiency?[13]

Indeed, with respect to inherent characteristics of approaching the urban through digital methods, one may ask: whoever is framing the urban issue towards which computation is addressed, are those groups nevertheless implementing a conception of the urban 'as just assemblies of functions and processes and of human agency as no more than the enactment of routines and procedures that can be embedded in software'? (Kallinikos 2011, p. 47). Through the use of digital technologies, wide-scale data can be brought together in unprecedented ways to inform (or even automate) decisive actions in real time. Inserting smart-city platforms into cities may consequently privilege those forms of computational codified knowledge, and gives greater affordance to functional and instrumental processes above other forms of knowing and acting in the city, such as knowledge embodied through acts of neighbourliness, or culturally informed ways of caring. Making a smart city operational would imply turning the actual city into a system:

> urban processes, agents and stakeholders, in order to be part of analytics, insights and action, have to be inside the presumptions of the software system itself … Whilst incorporating diversity and the ability to develop

> relationships, the need for modularity, interoperability and transferability across systems – and cities – revokes specificity.
>
> (Marvin and Luque-Ayala 2017, pp. 15–16)

On the other hand, data can be re-contextualised, and given greater social weight when framed and interpreted by the knowledges associated with plural experiences in the city (Tironi and Sánchez Criado 2015). Open and transparent computation might help reveal implicit political and economic interests residing in hitherto unquestioned or unreflective notions of system operational efficiency, functional integration and purposes like international competitiveness. Hidden assumptions in the codifications of the software, and the loss of situated knowledge in the abstractions of the platforms, can be identified, rendered explicitly and lead to corrections when code is open and programmes run transparently. According to Kurban *et al.* (2016, p. 14) and the idea of technopolitics, 'this dynamic and contentious process amongst various actors reconfigures political relations and power dynamics through conflicting appropriations as well as negotiations [of the technology]'. This would imply that the areas of urban life where computation is applied, and the agendas they serve, can be democratically controlled. Indeed, a critical challenge for further research focusing on the workings of smart-city systems in practice is to understand whether there are inherent characteristics of computational ways of knowing the city, how those play out in relation to non-computational ways of knowing and governing the city and whether digital projects can indeed be (re)conceived through urban politics that interact across elite and grassroots settings, creating more plural and hopeful possibilities, as is being attempted in Barcelona.

Acknowledgements

The research informing this chapter was funded by the Open Research Area collaboration between social science funding agencies in the UK (ESRC), Netherlands (NWO), France (ANR) and Germany (DFG). We are grateful to all the interviewees, contacts and colleagues in Barcelona who have helped with this ongoing study. This chapter presents preliminary findings, whose interpretation and presentation is the responsibility of the authors.

Notes

1 Copernicus Institute of Sustainable Development, Utrecht University. Corresponding author (e.dehoop@uu.nl).
2 Science Policy Research Unit (SPRU), Sussex University and Centro de Innovación Tecnológica para el Desarrollo Humano (itdUPM), Universidad Politécnica de Madrid.
3 Copernicus Institute of Sustainable Development, Utrecht University.
4 Urban Institute, University of Sheffield.
5 Urban Institute, University of Sheffield.
6 Copernicus Institute of Sustainable Development, Utrecht University and Monash Sustainable Development Institute, Monash University.

7 www.youtube.com/watch?v=OYNaoMkY8qY&feature=youtu.be (posted 15 July 2014; accessed 30 September 2016).
8 See the listing of multiple smart cities project at the Ajuntament website smartcity.bcn.cat (accessed 18 October 2016).
9 Whose neglect had provided space for artists and creative collectives now being displaced by the room being made for knowledge industry (Martí-Costa and Pradel i Miquel, 2012). A local campaign to resist and moderate 22@ to local needs had limited success.
10 The code and services offered by Decidim.Barcelona, and the Decide platforms from which it forked, are being taken up by other cities around the world and competing against commercial citizen participation services.
11 See, for example, projects funded by the Collective Awareness Platforms for Sustainability and Social Innovation funded by the EU H2020 and FP7 programmes. https://capssi.eu/.
12 Such studies are under way in the research project from which this more general case study chapter comes – see Knowledge Politics of Smart Urbanism https://smartknowledgepolitics.com/.
13 Initiatives like the Open Source Circular Economy Days are one forum for exploring these possibilities. https://oscedays.org/.

References

Aibar, E. and Bijker, W. E., 1997. Constructing a city: the Cerdà plan for the extension of Barcelona. *Science, Technology, & Human Values*, 22, 3–30.

Bakici, T., Almirall, E. and Wareham, J., 2013. A Smart City initiative: the case of Barcelona. *Journal of the Knowledge Economy*, 4 (2), 135–148.

Barcelona City Council, 2016. *Barcelona digital plan, 2017–2020: Transition to technological sovereignty*. Barcelona: Commission for Technology and Digital Innovation.

Capdevila, I. and Zarlenga, M.I., 2015. Smart City or smart citizens? The Barcelona case. *Journal of Strategy and Management*, 8 (3), 266–282.

Charnock, G., Purcell, T.F. and Ribera-Fumaz, R., 2014. City of rents: the limits to the Barcelona model of urban competitiveness. *International Journal of Urban and Regional Research*, 38, 198–217.

Continente, M., Pascual, J., Tomàs, M., Fiori, M., González, P. and Sunver, R., 2016. *¿Cómo pueden ser más inteligentes las ciudades?* Barcelona: Universitat Oberta de Catalunya.

Degen, M. and García, M., 2012. The transformation of the 'Barcelona Model': an analysis of culture, urban regeneration and governance. *International Journal of Urban and Regional Research*, 36, 1022–1038.

De Weerdt, J. and García, M., 2015. Housing crisis: the Platform of Mortgage Victims (PAH) movement in Barcelona and innovations in governance. *Journal of Housing and the Built Environment*, 31, 471–493.

Diez, T., 2012. Personal fabrication: fab labs as platforms for citizen-based innovation, from microcontrollers to cities. *Nexus Network Journal*, 14 (3), 457–468.

Eizaguirre, S., Pradel-Miquel, M. and García, M., 2017. Citizenship practices and democratic governance: 'Barcelona en Comú' as an urban citizenship confluence promoting a new policy agenda. *Citizenship Studies*, 1025, 1–15.

Estalella, A., Rocha, J. and Lafuente, A., 2013. Laboratorios de procomún: experimentación, recursividad y activismo. *Teknokultura*, 10 (1), 21–48.

Feenberg, A., 1999. *Questioning technology*. London: Routledge.

Gabrys, J., 2014. Programming environments: environmentality and citizen sensing in the smart city. *Environment and Planning D*, 32 (1), 30–48.

García, M., 2010. The breakdown of the Spanish urban growth model: social and territorial effects of the global crisis. *International Journal of Urban and Regional Research*, 34, 967–980.

Gonick, S., 2016. Indignation and inclusion: activism, difference, and emergent urban politics in postcrash Madrid. *Environment and Planning D: Society and Space*, 34, 209–226.

Greenfield, A., 2013. *Against the smart city*. New York: Do Projects.

Guallart, V., 2014. *The self-sufficient city*. Barcelona: Actar Publishers.

Gutiérrez, B., 2017. *Pasado Mañana*. Barcelona: Arpa Editores.

Hajer, M.A. and Dassen, T., 2014. *Smart about cities: Visualising the challenge for 21st century urbanism*. Rotterdam: NAI.

Jasanoff, S., 2004. *States of knowledge: The co-production of science and the social order*. London: Routledge.

Jordan, T., 2008. *Hacking*. Cambridge: Polity.

Kahn, R. and Kellner, D., 2007. Globalization, technopolitics, and radical democracy. In L. Dahlberg and E. Siapera, eds. *Radical democracy and the internet: Interrogating theory and practice*. New York: Palgrave Macmillan, pp. 17–36.

Kallinikos, J., 2011. *Governing through technology: Information artefacts and social practice*. Basingstoke: Palgrave Macmillan.

Kemp, R., Schot, J. and Hoogma, R., 1998. Regime shifts to sustainability through processes of niche formation: the approach of strategic niche management. *Technology Analysis & Strategic Management*, 10 (2), 175–198.

Kitchin, R., 2014. Making sense of smart cities: addressing present shortcomings. *Cambridge Journal of Regions, Economy and Society*, 8 (1), 131–136.

Kurban, C., Peña-Lopez, I. and Haberera, M., 2017. What is technopolitics? A conceptual scheme for understanding politics in the digital age. *Revista de Internet, Derecho y Política*, 24, 3–20.

Kuyper, T., 2016. *Smart City strategy & upscaling: comparing Barcelona and Amsterdam*. Barcelona: Universitat Pempeu Fabra.

Latour, B., 1991. Technology is society made durable. In J. Law, ed. *A sociology of monsters: Essays in power, technology and domination*. London: Routledge, pp. 103–131.

Latour, B., 2005. *Re-assembling the social – an introduction to actor-network theory*. Oxford: Oxford University Press.

Leon, N., 2008. Attract and connect: the 22@Barcelona innovation district and the internationalisation of Barcelona business. *Innovation: Management, Policy & Practice*, 10, 235–246.

March, H. and Ribera-Fumaz, R., 2014. Smart contradictions: the politics of making Barcelona a self-sufficient city. *European Urban and Regional Studies*, 1–15. doi:10.1177/0969776414554488.

Martí-Costa, M. and Pradel i Miquel, M., 2012. The knowledge city against urban creativity? Artists' workshops and urban regeneration in Barcelona. *European Urban and Regional Studies*, 19 (1), 92–108.

Marvin, S. and Luque-Ayala, A., 2017. Urban operating systems : Diagramming the city. *International Journal of Urban and Regional Research*, 41, 84–103.

Marvin, S., Luque-Ayala, A. and McFarlane, C., 2015. *Smart urbanism: Utopian vision or false dawn?* New York: Routledge.

Matthewman, M., 2011. *Technology and social theory*. New York: Palgrave Macmillan.

McDonogh, G.W., 2011. Learning from Barcelona: discourse, power and praxis in the sustainable city. *City & Society*, 23, 135–153.

Monterde, A., Calleja-López, A., Aguilera, M., Barandiaran, X. E. and Postill, J., 2015. Multitudinous identities: a qualitative and network analysis of the 15M collective identity. *Information, Communication & Society*, 18 (8), 930–950.

Ribera Fumaz, R., 2017. El modelo Barcelona, el urbanismo global y la teória urbana crítica. *Blogpost de TURBA* 1–12. Available at: https://turbain3.wordpress.com/2017/03/15/el-modelo-barcelona-el-urbanismo-global-y-la-teoria-urbana-critica/.

Romanos, E., 2013. Evictions, petitions and escraches: contentious housing in austerity Spain. *Social Movement Studies*, 2837, 1–7.

Smith, A., 2015. Tooling-up: civic visions, fablabs and grassroots activism. *Guardian*. Available at: www.theguardian.com/science/political-science/2015/apr/04/tooling-up-civic-visions-fablabs-and-grassroots-activism [accessed 4 October 2016].

Swyngedouw, E., 2009. The antinomies of the postpolitical city: in search of a democratic politics of environmental production. *International Journal of Urban and Regional Research*, 33 (3), 601–620.

Tironi, M. and Sánchez Criado, T., 2015. Of sensors and sensitivities towards a cosmopolitics of 'Smart Cities'? *Tecnoscienza*, 6 (1), 89–108.

Vanolo, A., 2016. Is there anybody out there? The place and role of citizens in tomorrow's smart cities. *Futures*, 82, 26–36.

Walt, V., 2015. Barcelona: the most wired city in the world. *Fortune*. 29 July. Available at: http://fortune.com/2015/07/29/barcelona-wired-city/ [accessed 16 October 2017].

York, R. and Rosa, E.A., 2003. Key challenges to ecological modernization theory. *Organization & Environment*, 16 (3), 273–288.

4 The professional knowledge politics of urban transport transitions in the greater Copenhagen region

Jens Stissing Jensen[1] *and Ulrik Jørgensen*[2]

Introduction

At its very core professional urban planning is about controlling and organising the complexity of the urban fabric. A fundamental prerequisite to such endeavours is the ability to translate state-of-affairs that for the average lay person appear chaotic and ambiguous into fact-like objects that are ordered and structured enough to enable interventions to be identified and prioritised. In order to render the urban fabric visible in the guise of well-defined and governable objects, engineers and other professional city planners employ carefully aligned analytical devices and epistemic frames, including models, calculations and statistics (see, for example, Chapter 1). In the hands of skilled professionals, these devices and frames make up powerful knowledge assemblages that are instrumental in transforming and purifying complex situations and processes into fact-like representations (Jørgensen 2012) that allow them to address urban reality as structured, predictable and manageable. The devices and epistemic frames used to translate complex and ambiguous situations into structured and knowable planning objects enables planners to construct and control planning rationalities that specify which objectives, concerns and arguments count when solutions and interventions are identified and prioritised in the practical production of urban reality. The stabilisation of these knowledge assemblages is often facilitated by institutional support in the form of specific authorised offices, spokespersons and standards that are instrumental in maintaining and reproducing centres of calculation (Latour 1999).

In some situations, the way in which the order of things is made intelligible by means of a specific professional-knowledge assemblage may, however, be challenged (Smith and Stirling 2007). Citizens may object to the professional justifications behind a project or an intervention based on their political convictions or everyday experiences. Such objections may draw attention to elements or experiences of urban reality that have been rendered invisible and irrelevant by a professionally controlled knowledge assemblage. In other cases, alternative planning communities may try to take control over the situation by rendering the order of things intelligible by means of alternative knowledge assemblages. In this sense transformations of urban planning rationalities are often the effect of controversies where different professional and political interests try to gain authority over how to render the order of things intelligible (Jørgensen 2012).

In this chapter we argue that such knowledge politics have salient implications for how planning rationalities pertaining to urban systems are developed and transformed. To illustrate this, we investigate how transport planning in Copenhagen has been influenced by a series of controversies over how professionals represent the order of things. A core issue in our discussion of such knowledge politics is how system boundaries and effects are defined and valued, not only as costs and benefits in classic economic terms, but also more broadly as effects related to trade, daily life, safety, cityscapes and health.

We illustrate our argument using four cases of knowledge politics pertaining to transport planning in Copenhagen since the Second World War. The first two cases illustrate how planning communities often cultivate knowledge assemblages that allow them to optimise individual subsystems independent of the outside world. The cases then investigate how planning rationalities pertaining to isolated subsystems may be transformed when actors make visible concerns that have been marginalised by dominant knowledge assemblages. The third case offers an example of how professionals may also employ more dynamic knowledge strategies in order to promote a particular subsystem by constructing and promoting multiple representations of the same subsystem. The fourth case provides an example of how transport planners may sometimes have to promote new planning rationalities in the vocabulary of knowledge assemblages imposed from the outside.

Case 1: the controversy over the Lake Ring Road in Copenhagen

This first case focuses on a planning controversy over car-based mobility infrastructure that has had fundamental consequences for the development of the transport system in the Greater Copenhagen Region since the 1970s. The controversy revolved around plans to construct a highway along the lakes surrounding the historic city centre. The rationalities involved in this controversy proved to be so contradictory that the role of car-based mobility infrastructure in Copenhagen remained unresolved for decades and still has ‘sleeping’ elements, for example highway projects lying in the drawers of transport planners in the Copenhagen municipality and the national government.

The backdrop to this controversy was a plan for the city’s expansion developed after the Second World War called the ‘finger plan’ (Egnsplankontoret 1947). Though it was never formally endorsed, this plan functioned as the modus operandi for most of the planning activities in the post-Second World War decades. This plan reflected a knowledge assemblage that associated spatial development of the Copenhagen Region (in terms of location of housing, industry and green areas) with the development of public transport infrastructures. More specifically, the finger plan outlined a regional development strategy where local trains would form trunk lines connecting the suburbs located along a set of 'fingers' stretching out from the city centre (Gaardmand 1994). The intention of the plan was to preserve large green areas between the 'fingers' and thereby provide dense city spaces with access to nature. The transport focus of the finger plan was the so-called S-train lines (essentially providing connections to the suburbs and surrounding areas) and trams that later were complemented by buses that connected the different parts of the Copenhagen region. This resulted in a strong emphasis on commuting between fingers and the city centre, with only a few transport lines connecting the suburbs in the different fingers. An important step in this process was the report from the government-supported Traffic Economic Committee that in 1961 presented its results, which argued for the rapid extension of the S-train network (Lyager 1996).

However, the finger plan and subsequent transport planning had not anticipated the huge growth in private car ownership that accelerated during the 1960s and resulted in fast-growing loads on interconnecting roads inside Denmark. This growth also resulted in increased car density in the historic city centre of Copenhagen, where roads were narrow and not designed for transportation dominated by cars. While the creation of a pedestrian road in the city centre reflected the visions of keeping cars out of the centre of Copenhagen, overall vehicle traffic in the city grew. This growth in car traffic was also poorly supported by the feeder roads into Copenhagen, which did not have separate lanes for each traffic direction, as the German motorways from the pre-war period did. A result of this was the establishment of a working group in Copenhagen with the specific purpose of planning a road connection from the Lyngby highway that connected Copenhagen with the northern parts of Zealand. The specific idea was to build a feeder road leading from the highway to the Copenhagen lakes by taking space from a public park and removing a block of houses along the road 'Fredensgade' (Lyager 1996).

While the growing importance of car ownership was essentially invisible within the knowledge assemblage underpinning the finger plan, it contributed to strengthening the influence of an alternative transport planning community established within the national governmental administration in 1949. This planning community was anchored in a governmental agency named the Road Directorate, which was established to take responsibility for national road and highway planning. Their position was strengthened by legislation in 1964 making them take over the planning of all major roads connecting cities, including main roads out of Copenhagen that earlier had been governed by the Copenhagen Municipality (Lyager 1996). This community cultivated car traffic as a discrete object of knowledge and expertise. New types of technical knowledge regarding road construction for motorised transportation was develop by a Governmental 'Road Laboratory' (Christensen 2003). This technical knowledge was complemented by the cultivation of expertise in socio-economic appraisal of road investments (Lyager 1996). With exclusive focus on car-based transportation, this community started to promote the extension of existing main roads and to initiate the building of highways, following German and

American principles. These plans came to dominate national traffic planning throughout the 1960s.

When the new highways approached the city boundaries of Copenhagen, cooperation between Copenhagen Municipality and the Road Directorate resulted in the planning of a system of ring roads with the purpose of diverting car traffic from the city centre. This included establishing a highway (Ring 1) along the lakes surrounding the centre of Copenhagen. This first general plan to include these highways was approved by the Municipal Council in 1958 without any protest. In addition to the planned ring road and its feeder roads, engineers in the municipality worked on a plan implying a complete rebuilding of a part of Copenhagen west of the main station, including the meat and hay markets (Københavns Kommune 1968).

In the late 1960s, the first road constructions connecting a highway to the city centre by a feeder road was initiated. This involved the construction of a tunnel and an elevated piece of highway, known as Bispeeng-buen, to circumvent some of the existing local streets. This project initiated a public controversy which was manifested in a public petition/protest signed by close to 50,000 citizens. A determining confrontation concerning the planned highway along the lakes (Ring 1) came in 1972, when squatters occupied a road that was planned to connect a motorway trunk line with the planned ring road. The squatters occupied the road for more than two weeks, despite the extensive use of the police force in an attempt to remove them (ironically, the road's name is Fredensgade, 'Peace Road'). The police eventually removed the protesters from the Peace Road, but as a result, Copenhagen municipality gave up constructing the last part of the feeder road connecting the Lyngby highway to the city centre and the planned 'Ring 1' road. Visitors to Copenhagen may note that even today this 'Ring 1' still does not exist. Today, a public park exists alongside the planned location for the feeder road.

The controversies surrounding the construction of Bispeengbuen, Ring 1 and the wider system of highways in Copenhagen resulted in a political stalemate that for the next two decades put new, large-scale traffic plans on hold. New concepts and planning criteria were needed, but could not be produced within the dominant knowledge frames.

Two professionally grounded knowledge assemblages clashed in this controversy. One had the national car infrastructure serving an all-purpose means of transportation – the car – as its focal visibility. This assemblage assigned value to the time spent on transportation between the regions and cities of Denmark. This assemblage was also linked to international highway connections followed by a series of bridge-building projects servicing this highway infrastructure and also related to the choice of highways to ensure the reduction of traffic accidents. The other assemblage valued the city as a living and working space connected by public transport systems. This assemblage rendered cars visible as a thread that would potentially become too dominant and space-demanding, resulting in a breakdown in the finger structure of the city. Consequently, the focus was on supporting public transportation and bicycling as the preferred means of transportation. This way of rendering the transport system visible was advocated by civil society groups and was reflected in the professional community of city planners as well. The case illustrates that the power relations among knowledge assemblages may be very difficult to anticipate. Until the confrontations over the construction of urban highways in Copenhagen, the car-focused knowledge assemblage appeared to constitute a hegemonic force, representative of societal modernisation and progress. Surprisingly, this highway and car-centred knowledge assemblage did not succeed in further promoting their planning rationality within Copenhagen even though they had strong advocates in the municipality's road-planning department. Faced with resistance from civil society resistance groups, weakly linked to powerful political parties and the city planning authorities having different priorities for the ideal traffic in the city centre, they could pursue their professional visions.

Despite the resistance and the avoidance of further elevated roads and wide ring roads in the centre of Copenhagen, the growth in car ownership has continued and consequently car congestion in the city centre remains problematic. In the municipalities surrounding Copenhagen, the transport planning rationalities are dominated by the idea of an omnipotent car system, which has resulted in widening of the outer ring roads. Also, plans to construct an underground tunnel highway northeast of the city centre has been strongly promoted in the recent decade. A first piece of connecting highway has been constructed without much protest compared to the case of the Bispeengbuen, not least because the combination of a tunnel and the use of an area formerly used for train maintenance has not spurred citizen protests. Still, 'Ring 1' has not reappeared as a grand solution and the municipal planners have adopted visions of 'liveability' as an alternative to car-focused planning, which reinvigorates former visions of Copenhagen as a city of cycling.

Case 2: multi-modality as transport rationale

Often societal functions such as transportation tend to be rendered visible as a series of independent and poorly coordinated planning objects (like, for example, cars, trains, cycling) by urban planners. The governance of compartmentalised subsystems can in periods overshadow options of planning for integrated multimodal solutions to addressing societal needs. In the post-Second World War decades, train services in Copenhagen were rendered visible as an increasingly independent planning object, and the planning rationality became isolated from the surrounding transport system. This case explores how the isolated knowledge assemblage informing railroad planning was successfully subjected to pressures for promoting trips that combined trains and cycling during the 1970s. This case thus illustrates how railroad governance moved from being centred on trains as an isolated and independent means of transportation to a new approach addressing trains in combination with other means of transportation.

For a century, the bicycle has been an important means of local transport in Denmark. Cycling reached the peak of its penetration as the dominant means of transport in Copenhagen around 1950, when bicycles shared road space with just trams and a few cars. Over longer distances bicycles were complementing travel by trains, which is reflexed in the fact that bicycles were allowed on trains in the early twentieth century (DSB 1942). From the 1960s the car became the dominant focus of traffic planning, as described in the previous case. This introduced competition between cars, bicycling and public transportation in the larger cities and led to a situation where the Danish State Railways (DSB) – the state-owned train operating company in Denmark – decided that bicycles could no longer be seen as a primary means of transportation. Bicycles were therefore not to be handled as travel luggage, but were managed as other goods.

By the early 1980s, bicycle traffic had increased by 50 per cent from its lowest levels in the late 1960s. Arguments promoting the use of bicycles were developed by the bicycle organisations and city planners (Krag 1985), and the bicycle was presented as one of the most promising contributions to solving the growing problems in society concerning energy, environment and the high costs of road extensions. Authorities also changed their concept of the bicycle (Rørbech 1997). Establishing better conditions for cyclists was no longer considered in terms of a matter of solving problems of traffic safety for a marginal group, but its forward planning also included environmental and economic concerns as well as concern over traffic capacity. Bicycle traffic once again became a discrete object of transport planning, and, since 1981, every public budget has improved conditions for cyclists by establishing bicycle lanes. The Ministry of Transport emphasised that the bicycle did not have to be 'reinvented' after this period of decline, as it had continued to be a visible part of traffic (Transportministeriet 1993, pp. 6–7).

After bicycles had been excluded for several decades from being taken on trains together with the bicyclist passenger, a major shift occurred in 1983 (Andresen 1981). The Danish State Railways was forced to respond to the new policy focus. Along with the change to the winter schedule the same year, the Danish State Railways chose to allow bicycles to be taken on the express trains to and from the largest Danish cities (up to a maximum of eight bicycles per train). Even though the Danish State Railways was reluctantly forced to take bicycles on trains, they changed both attitude and policy after this. This change in rationality was linked to the formation of a new, more cycle-friendly knowledge assemblage. Central to this knowledge assemblage was the definition of the commuter – rather than specific transport technologies such as the bicycle – as the pivotal knowledge object informing planning activities, as reflected in an interview with a railroad planner:

> For the first time within public transport, we openly acknowledged that commuters were the main target group for public transport. If public transport cannot serve commuters, then it has no chance of serving anyone else, because commuters have very well defined travel need.... The commuter project defined all the basic strategies that are still followed for S-trains.... It was completely clear to us at that time that the combination of bicycle

> and train was brilliant. The funny thing is that our customers had figured that out a long time ago, long before Danish State Railways did.
>
> (Kjærgaard 1999)

As planning began to centre on 'the commuter', cyclists became a central part of the business strategy of the Danish State Railways. This was because planners realised that more money was earned on a bicycle–train customer than on a bus–train customer: at the time, earnings from the total of public transport in the business strategy of the 'Danish State railways metropolitan area' were pooled and distributed on the basis of traffic counts.

The commuter- and cyclist-centred knowledge assemblage even influenced the physical design of the S-trains themselves. At the end of the 1980s it was decided that there was no train design on the market that satisfied Danish State Railways' wishes, which included demands for more seats, increased security, and fast boarding and exiting. A new S-train had to be designed. For this purpose, an internal idea group was established, who proposed the following design:

> A coach that was half as long as its predecessor, technically allowing an increase in width of 60 cm. The design ... meant two important things: We could have one more seat across [the carriage's width] and thus more seating capacity.... The other advantage was that we invented the concept of flexible areas inside the train that was large enough to be used for baby carriages [i.e. prams], wheel chairs, and bicycles.
>
> (Kjærgaard 1999)

The new design met DSB's formal demands, and, at the same time, it allowed the introduction of a storage space that significantly improved possibilities for taking bicycles on the commuter train (DSB 1995).

This case illustrates how urban governance of a societal function such as transport may become compartmentalised when planning communities are allowed to developed knowledge assemblage that narrowly focuses on optimising individual subsystems. The case, however, also illustrates that the boundaries needed to separate an individual subsystem from the wider transport system may be precarious and contested. In this case the formation of a new commuter-oriented knowledge assemblage proved instrumental in rendering the integration of cycling and train services visible as a relevant and valuable object of planning. This multi-modal rationality even become institutionalised in the physical design of S-trains in Copenhagen.

Case 3: cycling in Copenhagen – a flexible knowledge assemblage

The two cases above exemplify how planning communities often tend to construct knowledge assemblages that renders subsystems – such as highways or railways – intelligible as discrete planning objects that are to be planned and optimised largely independent of their contextual interdependencies with other systems or specific places. This case, which focuses on cycle governance, however, exemplifies how planning communities may also succeed in cultivating a more flexible knowledge assemblage that combines multiple way of knowing a given subsystem (Jensen *et al.* 2017).

The development of a flexible epistemic repertoire for making urban cycling visible as a governable phenomenon originated from a planning rationality informed by a knowledge assemblage making cycling visible as a safety hazard. This knowledge assemblage was originally developed in the 1930s where traffic accidents became a major concern due to the introduction of motorised vehicles in cities and consisted of accident statistics compiled from cycle accidents recorded in police reports. Key knowledge components of this assemblage included so-called 'black spots' and 'grey stretches', signifying places characterised by a higher-than-average frequency of accidents. Until the mid-1990s this knowledge assemblage was the basis of the municipal planning political rationality in relation to cycling in Copenhagen (Vejdirektoratet 1988).

From the early 1990s onwards, local politicians in Copenhagen began to request planning initiatives that actively promoted cycling. This policy pressure for 'something more' provided leeway for attempts to cultivate alternative ways of knowing and valuing cycling among planners in the municipal administration (Jensen *et al.* 2017). As an input to a cycle conference in 1996, a new analytical instrument for knowing urban cycling called the bicycle account (Københavns Kommune 2017) was developed and presented by planners within the municipal administration. The bicycle account is an accounting procedure that takes stock of cycling in Copenhagen on a biannual basis and, since 2002, the numbers presented in this account have constituted the basis for the formulation of planning objectives in relation to cycling in municipal strategies (Københavns Kommune, Bygge- og Teknikforvaltningen 2002). A key number in the account is the ratio of commuter trips taking place by bike. This ratio is assumed to be contingent on the quality of the cycle experience. The account, hence, includes quantified data on feelings of safety while cycling, experiences about the convenience of the cycling, and citizens' perceptions of the impact of cycling on the quality of the urban environment. These quantified experience-based numbers, which are generated through telephone interviews, are complemented by numbers on cycle-parking facilities, kilometres of green cycle paths that allows cyclist to ride detached from roads, etc. The cycle account thus renders urban cycling visible as an experience to be optimised and as a vital component of urban liveability. The good cycle experience is not translated into an economised market value, but is rather valued as good in its own right and as a means to increase the ratio of cycle commuting, which is a main planning goal.

The introduction of cycling experiences as a new, quantified knowledge component within the established safety-oriented knowledge assemblage has promoted a wide variety of interventions in cycling infrastructure. More cycle paths have been constructed independent of car-based road infrastructure, and many new cycle bridges have been constructed over the harbour and over main feeder roads. In addition, a main street that runs through an urban district near the city centre has been transformed into a so-called cycle street, with broader cycle lanes and limited access for cars. Currently a main policy focus is parking for cycles. A key goal of a recent municipal planning strategy hence is to ensure that 70 per cent of the cyclists experience the parking facilities as 'good' or 'very good' (Københavns Kommune 2015). Based on this goal a detailed registration of established cycle-parking facilities has been initiated. This registration has shown, for example, that only 74,000 public cycle-parking facilities are currently available to serve a total number of some 680,000 cycles.

The introduction of cycling experience as a discrete knowledge component within the knowledge assemblage pertaining to urban cycling illustrates how new planning rationalities may be cultivated in the interplay between changing political signals and the development and institutionalisation of new analytical ways of knowing. The precondition for the development of the cycle account was thus political signals that a more active way of planning for cycling should be developed. The established accident-oriented knowledge assemblage was incapable of translating these political signals into an operational planning rationality since it did not ascribe a positive value to cycling. The cycle account helped render the translation of political signals into an operable planning rationality by providing a structured way of knowing cycling as a positive contribution to urban life. In this way planners in Copenhagen contributed to the development of what had been a relatively weak and marginalised planning object into a salient and operable component of urban development by cultivating a multi-pronged knowledge assemblage that renders cycling knowable both as a safety risk and as a valued urban experience. This knowledge assemblage has enabled cycling to be promoted within different governance arrangements concerned with, for example, safety and liveability. A disadvantage is that the various valuations embedded in the knowledge assemblage may occasionally run into conflict with each other. Strategies to change traffic regulation in support of more convenient cycle experiences may come into conflict with planning goals that have a narrow focus on accident. Multi-pronged knowledge assemblages may hence offer less coherent guidelines for system development and system-optimisation. An effective application of such planning rationalities hence requires planners that are competent in manoeuvring within such contradictions.

While street-level experimentation is often promoted as the main technology of governance in the transition literature (Schot and Geels 2008; Voytenko *et al.* 2016), this case suggests that the development of devices that are capable of institutionalising new ways of knowing is also of central importance when political signals are to be translated into longsighted and operable planning rationalities.

Case 4: navigating within a socio-economic knowledge assemblage

This final case – which also focuses on bicycling governance in Copenhagen – illustrates how urban planning communities sometimes need to frame their knowledge assemblages in terms of the dominant vocabularies dictated by other professional communities that have gained power. It also illustrates, however, that there may be ample opportunities for manoeuvring within such externally defined epistemic frames in some contexts.

The context for this case is a hegemonic knowledge assemblage that has been cultivated by the Ministry of Finance in Denmark over the preceding decades in order to enable the Ministry of Finance to take control over policy coordination across various areas of the national government administration. The main rationality of this knowledge assemblage is to optimise the economic competitiveness of the Danish industry by decreasing the growth in public expenditures and by increasing the supply of labour. In addition to supply-side-oriented macro-economic models, a main component of this knowledge assemblage is socio-economic analysis. Socio-economic analysis has been promoted as it enables the Ministry of Finance to calculate and compare the societal effects of investment and policies across different policy resorts, thus allowing for a more efficient allocation of resources (Finansministeriet 1999). Resource allocation thus increasingly depends on the ability of the different ministries to demonstrate their societal relevance in the language of socio-economic analysis.

Socio-economic analysis is not a precise knowledge device since both the effects to include and the valuation of these effects may be determined in very different ways. The methods applied to carry out socio-economic appraisals have thus become subject to intensified knowledge controversies; for example, in relation to the valuation of transport investments.

Traditionally, socio-economic analysis of transport investments only includes so-called direct effects. These effects include travel time, noise, pollution and accidents (Copenhagen Economics 2014). In 1999, the Ministry of Finance issued new guidelines that introduced so-called 'broader effects', which are effects that in one way or another are derived from the direct effects (Finansministeriet 1999). More specifically, the guidelines introduced so-called market distortion effects from tax-finance expenditures as a mandatory socio-economic component to be included in socio-economic calculations. Taxes are presumed to disincentivise people from working, thus having a negative effect on the national supply of labour. Based on this reasoning the new guidelines determined that 1 Danish Kronor (DK) of taxed financed expenditure represents a socio-economic cost of 1.2 DK, due to such market distortion effects. To some commentators this market distortion effect was seen as a political tool introduced to render public investments, financed through taxes, socio-economically less favourable than market-based investments (Bøndergaard 2012).

The introduction of market distortion effects illustrates that socio-economic calculations can be engineered to serve particular objectives, and consultants soon recognised this type of 'valuation engineering' could form the basis of a commercial service. Among these was a consultant named Copenhagen Economics. In 2014, the Ministry of Transport commissioned this consultant to identify 'broader socio-economic effects' of transport investments. The objective of this analysis was to identify broader socio-economic effects that would increase the calculated socio-economic benefit from transport infrastructures investments, thereby strengthening the bargaining position of The Ministry of Transport vis-à-vis the Ministry of Finance. In line with this objective, Copenhagen Economics identified a series of 'broader socio-economic effects' that made traffic infrastructure investments appear more socio-economically favourable (Copenhagen Economics 2014). They argued that transport investments alleviate a series of existing market-distortion effects. Commuting is hence portrayed as a market-distorting effect that disincentivise people from working. By reducing transport time, new transport investments are believed to reduce this disincentive, thus increasing the supply of labour in the economy. Using the same type of argument, new transport infrastructures are also argued to reduce market distortion effects in relation to markets for goods and services. Furthermore, new transport infrastructures are argued to generate so-called 'agglomeration effects'. This type of effects is argued to exist because spatial proximity among companies and labour in the same sector increases competition, learning, innovation and productivity. By increasing such proximity in particular sectors, new transport infrastructures are believed to produce positive socio-economic effects. The analysis thus concludes that the direct effects included in traditional socio-economic analysis represent a very incomplete picture of the actual societal benefits of new transport infrastructure.

City administrations are becoming increasingly aware of socio-economic analysis as a powerful way of knowing – especially in relation to attempts to influence the national governmental administration. In 2009 the municipal administration in Copenhagen introduced socio-economic calculation as a new knowledge device in relation to cycle planning (COWI 2009). This device took a point of departure in the standardised methods for socio-economic calculation developed by the Ministry of Transport that were designed to be applied to motorised road traffic and public transportation. This method was not passively adopted, however, but was actively re-engineered in support of cycle investments. The decisive difference compared to the standard socio-economic method was the inclusion of public health effects. According to the method developed by the municipality, health benefits amount to a socio-economic value of approximately 6 DK (0.8 euro) per kilometre cycled due to better health and prolonged lifetime. The calculation thereby frames cycling not only as a relevant means of transportation, but also as the single most important element of public health promotion in the city. This way of re-engineering the socio-economic calculation of cycle investment had tangible effects on the governance of cycling. After the method was developed in 2009, public health promotion was introduced in the socio-economic manuals of the Ministry of Transport (Transportministeriet 2013). The method was thereby instrumental in rendering cycling as a visible and attractive mode of mobility in national policy formulation. This provoked the Ministry of Transport and regional authorities to co-finance a new type of inter-municipal cycling infrastructure that aims to support cycle commuting over longer distances, a so-called cycle super highway (Sekretariatet for Supercykelstier 2017).

The case illustrates that urban planning communities often do not have complete autonomy over their own knowledge assemblages and the resulting planning rationalities pertaining to their area of planning. Urban planning communities, therefore, may need to articulate their planning rationales by means of knowledge assemblages shaped and partly controlled by other actors. However, the case also illustrates that urban planning communities may creatively construct planning rationalities in the vocabulary of externally defined knowledge assemblage for their own purposes.

Discussion and conclusions

Urban governance of complex societal functions such as transport is an activity seldom carried out in an entirely coordinated and integrated way, not least because such coordination would demand an overarching perspective – a privileged 'view from nowhere'. Instead, such functions are typically divided into a series of partial planning objects related, for example, to cars, railways and cycling. The governance of these planning objects is then taken up by particular groups of actors and planning communities. In this chapter, we have illustrated how such planning communities cultivate knowledge assemblages in order to construct and control planning rationalities pertaining to individual subsystems within the transport system. We also demonstrate how continuous work is invested in maintaining, contesting and modulating such knowledge assemblage.

The knowledge assemblages and planning rationales of professional planning communities are thus typically subject to a complex series of stabilising and destabilising pressures. On the one hand, members of established planning communities typically have good professional reasons for stabilising established knowledge assemblages. A central concern is to maintain control over the predictability of particular planning objects. To this end professional planning communities may try to make individual subcomponents of the transport system (such as cars and train) knowable and governable as discrete planning objects, that can be dealt with in isolation from other components of the transport system, and independently of the specific contexts in which the transport systems operate. The advantage of isolating subcomponents from the outside world is that it allows for a more operational and clearly defined understanding of how to optimise the system. When established knowledge assemblages lose control over planning rationalities pertaining to a specific planning object there is a risk that the planning object becomes ungovernable. This has been the case for road planning in Copenhagen, for example, since a specific highway-oriented knowledge assemblage lost control over planning rationalities in the 1970s, as discussed in case one.

On the other hand, established knowledge assemblages are typically also subject to pressure for change, especially if these assemblages are experienced as too myopic and sub-optimising by outside actors and interests. A salient concern in transport planning in Copenhagen since the 1970s has thus been to promote planning rationalities that ensure that individual subsystems are sufficiently coordinated with the development of the broader transport system and the nature of the specific contexts in which the system is introduced. Both cases one and two illustrate how knowledge assemblages characterised by narrow system boundaries have proven difficult to maintain when these assemblages do not accommodate integrated and place-sensitive planning rationalities.

As also suggested by cases one and two, professional planners may sometimes be reluctant to redefine system boundaries. In other situations, planners have actively engaged in the construction and promotion of knowledge devices that redefine system boundaries in order to strengthen extant planning objects. Cases three and four thus illustrated how cycling has been developed into a central urban planning object through the cultivation of knowledge devices that systematically relate cycling to urban experiences and the production of public health. In addition, case four illustrates how urban knowledge production may be influenced by externally defined knowledge assemblages. Over recent decades engineering-based knowledge assemblages focusing on optimal system performance have increasingly been recast in the language of socio-economic analysis. The cases illustrate that urban planners have taken an active part in engineering this vocabulary to their own advantage.

More generally, the cases illustrate that political rationalities pertaining to urban systems are tightly related to professionally controlled knowledge assemblages that control everyday knowledge production. These assemblages construct the boundaries and relations through which the city becomes visible as well-defined and governable planning objects. The strategies to transform established political rationalities and institutionalise new ones by politicising and reconfiguring the 'nuts and bolts' of extant knowledge assemblages remain underexplored within the transition governance literature.

Notes

1 Department of Planning, Aalborg University, Denmark.
2 Department of Planning, Aalborg University, Denmark.

References

Andresen, O., 1981. The train must leave on time [Der skal være afgang til tiden]. *Politiken*, 20 Dec.

Bøndergaard, M.R., 2012. Finansministeriets regnemetoder er tynget af politisk arvegods. *Information*, 10 Sep.

Christensen, J., 2003. Road technological institute then and now [Vejteknisk Institut før og nu]. *Dansk Vejtidsskrift*, 9–11.

Copenhagen Economics, 2014. Broader economic effects of transport investments [Bredere økonomiske effekter af transportinvesteringer].

COWI, 2009. *Economic evaluation of cycle projects – methodology and unit prices*. DSB, 1942. Bicycle transport 50 years ago [Cykelbefordring for 50 år siden]. DSB's internal personnel magazine *Vingehjulet*.

DSB, 1995. The S-train must get closer to the customers [S-toget skal tættere på kunderne]. *DSB magazine*, 10–11.

Egnsplankontoret, 1947. *Draft proposal for regional plan for Greater Copenhagen [Skitseforslag til Egnsplan for Storkøbenhavn]*.

Finansministeriet, 1999. *Guide to preparation of socio-economic effect evaluation [Vejledning i udarbejdelse af samfundsøkonomiske konsekvensvurderinger]*.

Gaardmand, A., 1994. The Finger Plan is a real common sense plan [Fingerplanen er en rigtig snusfornuftig plan]. *DSB magazine*, 3, 18–19.

Jensen, J.S., Cashmore, M. and Elle, M., 2017. Reinventing the bicycle: how calculative practices shape urban environmental governance. *Environmental Politics*, 26 (3), 459–479.

Jørgensen, U., 2012. Mapping and navigating transitions – the multi-level perspective compared with arenas of development. *Research Policy*, 41 (6), 996–1010.

Kjærgaard, E., 1999. Interview made by the INTERPOL PROJECT.

Københavns Kommune, 1968. *City Vest 1968 – traffic- and Center Plan for inner Vesterbro in Copenhagen [City Vest 1968 – Trafik- og centerplan for det indre Vesterbro i København]*.

Københavns Kommune, 2015. *Community Copenhagen [Fællesskab København]*.

Københavns Kommune, 2017. *The Bicycle account 2016 [Cykelregnskabet 2016]*. Teknik- og Miljøforvaltningen.

Københavns Kommune, Bygge- og Teknikforvaltningen, V.&P., 2002. *Cycle policy 2002–2012 [Cykelpolitik 2002–2012]*.

Krag, T., 1985. Danish Cyclist Federation 80 years [Dansk Cyklist Forbund 80 år]. *Cyklister*.

Latour, B., 1999. Circulating reference: sampling the soil in the Amazon forest. In B. Latour, ed. *Pandora's hope: Essays on the reality of science studies*. Harvard, MA.: Harvard University Press, pp. 24–79.

Lyager, P., 1996. *Ten years with a general plan outline 1958–1968 [Ti år med en generalplanskitse 1958–1968] Byplanhistoriske noter 33*.

Rørbech, J., 1997. Introduction. In *Copenhagen – city of cyclists*. Københavns Kommune, Magistratens 4. afdeling.

Schot, J. and Geels, F.W., 2008. Strategic niche management and sustainable innovation journeys: theory, findings, research agenda, and policy. *Technology Analysis & Strategic Management*, 20 (5), 537–554.

Sekretariatet for Supercykelstier, 2017. *Supercycle highways 2017–2045 [Supercykelstier 2017–2045]*.

Smith, A. and Stirling, A., 2007. Moving outside or inside? Objectification and reflexivity in the governance of socio-technical systems. *Journal of Environmental Policy & Planning*, 9 (3–4), 351–373.

Transportministeriet, 1993. *The bicycle in Denmark : present use and future potential*.

Transportministeriet, 2013. *The socio-economic effects of cycling [Cyklingens effekter og samfundsøkonomi]*.

Vejdirektoratet, 1988. *Cycle paths in cities – the safety effect [Cykelstier I byer – den sikkerhedsmæssige effekt]*.

Voytenko, Y., McCormick, K., Evans, J. and Schliwa, G., 2016. Urban living labs for sustainability and low carbon cities in Europe: towards a research agenda. *Journal of Cleaner Production*, 123, 45–54.

5 The power-knowledge of best practice

Governing climate change in German municipalities

Nanja Christina Nagorny-Koring

Introduction

Imagine the following situation: the region of Hanover (Germany) has received funding from the German federal government to develop a climate-adaptation strategy. After one year of concept development, the strategy must be put into action. As in most municipalities, however, climate-change adaptation is a fairly new issue for the regional government. The increase in the number of invasive species due to climate change, biodiversity-enhancing concepts such as urban ecology, and ways to deal with increasing wind speeds and intense precipitation events are unchartered territory for municipal staff. Furthermore, climate policy is a voluntary task for local governments in Germany. Given that there is no legal requirement for them to have a climate policy, resources available for this work are limited. Thus, decision-makers and implementers need to carefully justify any climate-related actions (Benz *et al.* 2015). In these circumstances, cognitive, symbolic and material resources are required to aid the municipal staff in addressing the challenges. Questions facing the people in charge of municipal climate action are as follows: what actions are appropriate? How can any action be funded? How can the municipality satisfy relevant funding criteria? How can the initiatives be effectively pitched to the local governing body? What works and what does not? What mistakes are avoidable? Thus, these questions imply that experience and knowledge are of great importance when it comes to new challenges like urban climate governance.

In order to provide support to municipal staff and to accelerate the implementation of local climate action, a large number of best climate-practice examples have been produced by various governmental and non-governmental organisations (NGOs). Disseminated through handbooks, online databases, or conferences, best climate-practice descriptions depict a broad range of purportedly successful examples, including holistic climate-strategy design, citizen participation initiatives and individual (technical) measures (Bulkeley 2006; Stead 2012). Enhanced networking and exchange of ideas, and increasing sources of information, are assumed to increase the efficiency and efficacy of climate-change action, besides saving time, money and labour. Hence, the exchange of best practice at local, regional and national levels is considered an important approach to finding solutions to climate-change challenges. In this context, it has become received wisdom that the dissemination of best practice can lead to policy change, because a sufficient number of best-practice examples, which merely require adaptation to local circumstances, are said to be already available (Brand 2007; Schreurs 2008).

The promotion of best practice as a governing technique is not restricted to urban climate governance. No policy area appears to be exempt from the use of best practice: social justice policy, urban regeneration policy, spatial policy, regional development policy, environmental and climate policy, and urban sustainability policy are but a few examples of sectors for which best practice is said to be relevant. No other governing technique is more popular and widely diffused than best practice, it appears. It is used at all levels of government, ranging from communities to supranational entities, and across all scales and spatial contexts (Vettoretto 2009). Recommendations for best practice abound in:

- Academic literature (e.g. Demeritt and Nobert 2014; Schechtman 2015);
- Conferences: for example, the promotion of '10 best practices for sustainable urban development' (Fieldnotes-160601-Berlin);
- Transnational municipal networks (TMNs) for climate action that publish best-practice guidelines and online databases;
- Governments and supranational institutions such as the EU, whose *Urban Agenda* (EU 2016) focuses explicitly on financing pilot projects to be replicated EU-wide and for the exchange of lessons learnt among EU countries (Radaelli 2004; Vettoretto 2009; Tedesco 2010; Stead 2012; Boulanger and Nagorny 2018).

Despite an endless number of best-practice examples, little is known about best practice as a governmental technology (Bulkeley 2006; Macmillen and Stead 2014; Montero 2017). Examples of best practice are often uncritically and unreflectively accepted, and their production, reproduction, diffusion and application are often deemed quasi-natural in policy-making (Vettoretto 2009). As a result, best practice has become established as a key governmental technology with limited critical scrutiny. Examining best practices in municipal climate action provides interesting insights due to the relative novelty of this policy area, because best-practice examples are always assigned particular relevance in light of new circumstances and challenges, as illustrated by the example at the beginning.

In this chapter, I consider best practice conceptually as a powerful knowledge-configuring device that actively co-defines the nature of urban reality in line with the political-strategic interests of its 'producers' and 'users'. To do so, I apply a governmentality perspective – as outlined in Chapter 1 – to illustrate, first, the constitutive role of knowledge in urban climate governance in rendering the urban fabric visible in specific ways. Second, a governmentality perspective allows me to explore the role of epistemic devices in the production and enactment of authoritative knowledge. Third, it helps me to critically reflect on the suitability of governing by best practice to accelerate urban sustainability transitions.

To highlight the role of epistemic devices in shaping urban transitions, I will uncover the hidden production processes of best practices. More specifically, the chapter analyses how best-practice examples are selected and produced in the context of municipal climate action in Germany. This case study incorporates questions as to how contextual experiences and actions, as well as tacit knowledge, are addressed in the creation of best-practice standards, and what is intentionally removed or obscured to make a particular practice example appear as 'the best'.

The chapter consists of six sections. Following the introduction, I outline a knowledge-politics perspective on best practice. This theoretical section is followed by a presentation of the methodological approach. Fourth, I examine the power relations existing between the selection and production processes of best practice for urban climate-change action in Germany. Fifth, I discuss the role of best-practice guidance in configuring political rationalities of urban climate governance. The chapter ends with concluding remarks on the transformative potential of best practices for urban climate-change governance.

Theorising best practice

Despite best practice being a widely, and largely uncritically, accepted governing technique, it lacks a clear definition in both theory and practice. The following definition has been suggested (Liem and McInerney 2008, p. 3):

> Best practice refers to the most efficient and effective way of accomplishing a task, based on repeatable procedures that have proven themselves over time. It is efficient because they require minimal amount of resources (i.e. time and effort) and it is effective as they promise solid, evidence-based, and state-of-the-art work in a particular field that has proven to reliably lead to a desired outcome.

Best practice is accordingly understood as the best method or approach to accomplish a task. This suggests that best practice is about accumulating and applying specific knowledge regarding what does or does not work in a particular situation or context. This definition prioritises continuous learning, feedback, reflection and analysis surrounding the questions of how and why something works (Bendixsen and Guchteneire 2003).

But how are best practice examples generated, and which power relations unfold during this knowledge-creation process? What understanding of urban climate-change policy emerges from the interaction between political rationalities and the notion of best climate-practice? Foucault's concept of governmentality provides a critical perspective to deal with these questions. It helps to identify different mindsets, reveal their origin and analyse the principles and knowledges they rely on. It also provides insights into the embedded practices best practice engenders and how the embedded practices themselves help operationalise these rationalities (Rose *et al.* 2006).

Apart from his main topic of interest – subject formation – Foucault's intellectual work focuses extensively on the analysis of power relations, especially the interplay of power and knowledge. He assumes that power, in contrast to domination, is not repressive but rather enabling and productive (see Chapter 2). Foucault (1995) thus asserts that power and knowledge directly imply and reinforce each another. He suggests that mechanisms of power such as screenings, monitoring, evaluation, traditional identity checks and surveillance produce new, and increasingly detailed, knowledge, which in turn can be used to improve the efficacy of power mechanisms or to create new methods of control. In this context, Foucault refers to 'power–knowledge relations' (Foucault 1995, p. 27).

The capacity to generate knowledge about the environment, as a precursor to its governance, needs to be understood as the key to defining management mechanisms. Carbon footprinting, climate-change scenarios and pathways, and best practices contribute to the establishment of particular power–knowledge relations, whose characteristics can be measured, rated, evaluated and managed. Climate governance, therefore, co-evolves with the construction of particular climate truths and their diffusion using specific technologies, methods, practices and discourses. To develop visions, strategies and concepts for municipal realities, relevant actors depend on knowledge created through maps, statistics, models and best-practice examples. Consequently, municipal climate action is not a fixed entity awaiting description, but rather a governable phenomenon whose existence is constructed through interactions between power and knowledge. Forms of knowledge such as best practice should not be represented as neutral 'shortcuts' of municipal realities. Instead, it is necessary to bear in mind that best-practice examples reflect a selective and powerful interpretation of these realities. Hence, different analytical forms of knowledge, of which best practice is one, create different representations of municipal realities (see Chapter 1).

Researching best practice as a knowledge-producing governmental technology enables the reciprocities between these technologies and their political rationalities to be critically analysed by revealing the underlying power–knowledge complex. This is of particular relevance in the area of climate governance with its long time frames and transformative nature, wherein knowledge creation through epistemic devices can be understood, on the one hand, as an attempt to control politically contentious situations by describing interventions free from disruptions, problems and conflicts. On the other hand, knowledge-producing devices can create new realities, as opposed to stabilising extant visions, which may disrupt and ultimately change existing systems, thus promoting existing societal tensions in a bid for transformative change, instead of rendering contradictions invisible. Hence, from a normative perspective, it is important to ensure that the knowledge-producing processes, which support the formulation of political realities, are configured in a manner that not only increases the visibility of social transformations towards sustainability and climate-change action, but ultimately also makes them happen (see Chapter 1).

Thus, governmentality is a productive resource to enhance the 'think-ability and criticize-ability' of contemporary governance forms (Walters 2012, p. 2). Such a perspective provides new insights into local climate action as a political space by re-problematising established thought and behavioural patterns, which make specific governmental strategies, such as best practice, appear quasi-natural. In terms of urban climate governance, it enables us to better understand how governing climate change operates on the ground and 'to offer an explanation for how and why particular forms of authority, knowledge, and practice have come to dominate the climate change field' (Stripple and Bulkeley 2014, p. 14).

Tracing best practice in German local climate action

To explore these issues further, I use as an empirical case – the German funding programme 'Masterplan 100% Climate Protection'. As an 'excellence initiative', the Masterplan programme features high climate ambitions, aims at developing model solutions, and provides increased funding for action in this field. Between 2012 and 2018, 19 municipalities, each having a population of 7,000–700,000, were supported through this programme. The geographical distribution of these 19 municipalities is shown in Figure 5.1. In 2016, 21 additional municipalities joined the programme, but these have not been taken into consideration in this research. The Masterplan programme aims to draw public attention to, and stimulate the reproduction of, exemplary solutions to achieve 95 per cent emission and 50 per cent energy-demand reductions by 2050. The rationale behind choosing 40 municipalities as frontrunners for this programme lies in showcasing different decarbonisation pathways that can then be adopted by as many local authorities as possible (Ziesing 2010; Bundesministerium für Umwelt, Naturschutz, Bau und Reaktorsicherheit 2015).

I collected primary data through 35 in-depth interviews with 20 German climate managers, six representatives of TMNs,[1] five policy consultants, two municipal climate experts from research institutes, an NGO and an employee of the German Ministry of the Environment. The interviews were semi-structured, with only open-ended questions asked. In addition, I attended 20 conferences and workshops in Germany, where best practice cases were presented to experience how best practice unfolds in practice as an epistemic device. These events included the *International Climate Alliance Conference* (2015, Dresden), the *International Conference on Climate Action* (2015, Hanover), and nine networking meetings of German municipal climate managers (2015–2017).

Secondary data in the form of policy documents, such as municipal climate strategies and federal support programmes in the German context, as well as best-practice databases, guidelines and handbooks from both national and international sources were analysed to contextualise the interview responses. This

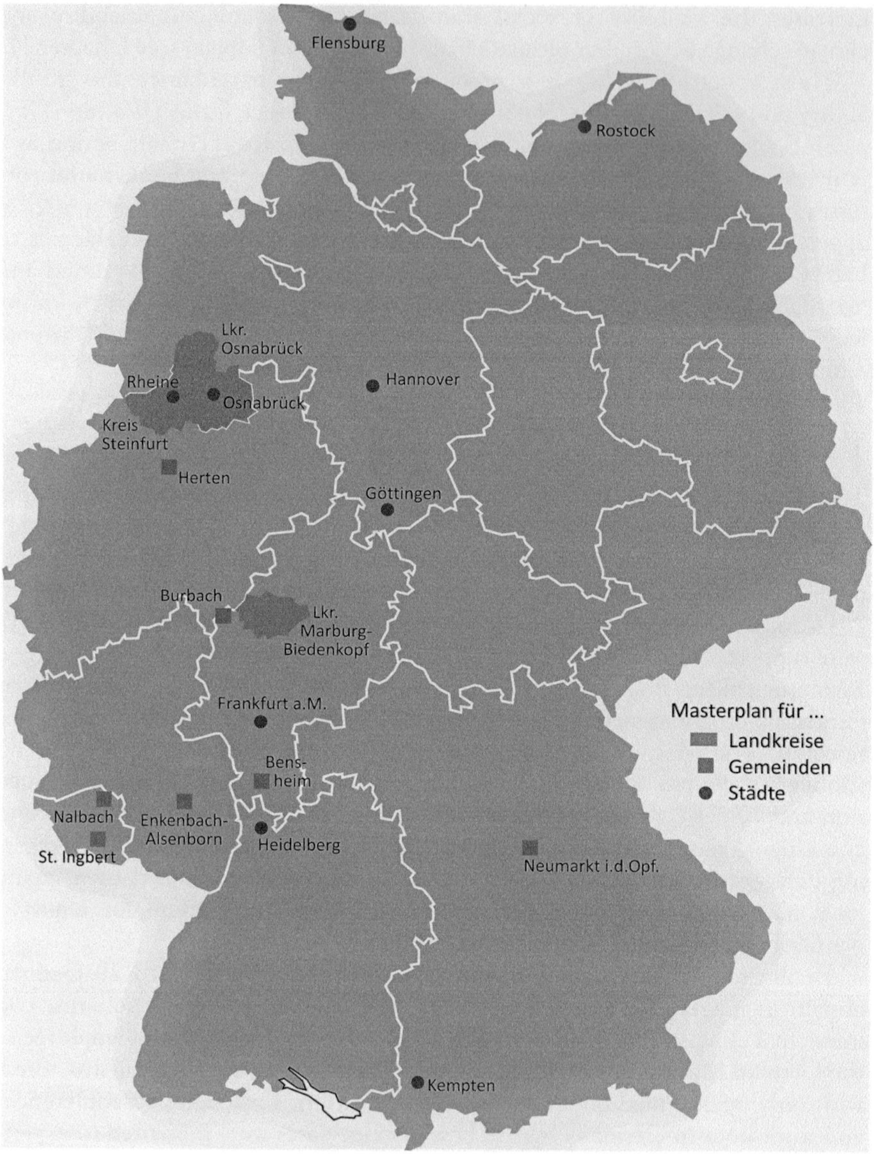

Figure 5.1 Geographical distribution of the studied Masterplan municipalities in Germany.

Source: map created by Elke Alban, Department of Human Geography, Goethe University Frankfurt/M.

led to a corpus of 368 descriptions of best practices which I analysed quantitatively. Key variables in the quantitative inquiry were:

a The geographical origin of the best-practice example;
b The characteristics of the example (i.e. technical, non-technical, socio-technical; holistic or single measure);
c Level of detail; and
d Discussion about transferability and barriers.

Since each method reveals a different aspect of empirical reality, methodical triangulation leads to a greater variance in the collected data (Patton 2002): First, participatory observations during conferences and networking meetings allowed me to understand the epistemic politics at work. Second, the qualitative interviews made the hidden practical functioning and power relations of best practices as epistemic devices, visible to me since I spoke to the users and producers of best practice. Third, the document analysis provided mainly representational data about the content of best practices.

Selection and production of best practice

What practices are identified as the best and most replicable? What criteria are used to select them? What actors are consulted or involved? Put simply, how is best practice for municipal climate action created out of everyday governance?

The aim of the best-practice approach is to inductively identify generalisable principles that can improve and streamline interventions and procedures through observation and evaluation of successful individual practices. This abstraction of many experiences across a wide variety of projects into instructions for action is thought to enable systemic comparisons either through a quantitative micro-economic or a qualitative case-study approach (Veselý 2011). In practice, the case-study approach dominates: individuals from various organisational contexts use databases, interviews, conferences, academic and grey literature, as well as the internet, to identify cases that can be considered 'best practice'. This involves decisions regarding what constitutes adequate actions and interventions, as well as the descriptions of replicable and desirable examples.

The selection process is more haphazard than appears at first sight. Producing best-practice examples is rarely a systematic endeavour; in most cases, it is linked to project requirements, thus often implying financial and temporal limitations established in the individual project design. A systematic qualitative or quantitative review of relevant examples from climate-active municipalities is impractical given the amount of activity occurring in this field: in Germany, between 2008 and 2016, more than 10,000 local climate projects in around 3,000 municipalities were supported as part of the national climate protection initiative (NKI) (Bundesministerium für Umwelt, Naturschutz, Bau und Reaktorsicherheit 2017). As Lindblom (1959) correctly identified, decision-making in public services, administration and ministries is not the product of quantitative or qualitative methodological analysis and comparison of optimal solutions. Instead, the governmental process follows a more pragmatic and less transparent approach. The same applies to the identification and preparation of best practice examples: the process is restricted to 'successive limited comparisons' (Lindblom 1959; Nagorny-Koring and Nochta 2017). My research on the case of the Masterplan programme revealed three types of emergent best practices: (1) those arising by chance; (2) self-runners; and (3) politically championed ones.

To begin with, people often stumble by chance across what they deem are good examples through personal conversations, conferences or from reading. Actors are sometimes already aware of some suitable candidates for potential best-practice examples due to existing contacts and information (Interview Network-5 2016). According to a municipal climate-change manager, the emergence of best practice 'has a lot to do with chance: who is where at what time' (Interview Municipality-15 2016).

Second, there are many practices that, due to their nature, become viewed as attractive without any intervention from external actors (Interview Municipality-4 2015). An example from the Masterplan programme is the 'Klimapakt Flensburg' (http://klimapakt-flensburg.de). The idea behind this programme was to outsource municipal climate protection to a private association to make local climate action independent of political influences and thus more stable. This idea spread quickly to other municipalities (Bensheim, Burbach and Kreis Steinfurt) and quickly became established as a best practice.

Third, political motivations are important in the selection and promotion of best-practice examples. Sometimes municipalities – especially those that want to foster a 'green' image – actively promote some of their interventions as 'best practice' by disseminating them through networks or conferences, submitting them to competitions, etc. As a municipal climate-change manager explains,

> [...] opportunities for diffusion are increased by turning up to as many events as possible, by presenting it, by diffusing it using city networks. These I believe are important aspects for increasing visibility [...] ultimately, it's about city marketing and what can be used for external communication.
>
> (Interview Municipality-15 2016)

What is critical then is the amount of resources spent on promoting a supposed example of best practice. Moreover, many best practices are developed in the context of third-party projects, because the production of best-practice guidance is often a generic project output, expected by funders. Often the funder has particular political interests, whereby specific thematic, spatial or methodological criteria are established, and examples of best practice are produced and disseminated based on these interests. The availability of best practices can hence be limited by thematic constraints, because the production of best practice then depends on how the availability of funding is framed (Interview Network-5 2016).

How a 'simple' practical example is elevated to be best climate practice also depends on the knowledge 'producer' or champion: a TMN, a research institute working for government, or a consultancy working for a municipality or a ministry. The interests and motivations leading to the identification and diffusion of best-practice examples can vary, as does the capacity of different groups to do so. What becomes best practice strongly depends on the effectiveness of an organisation in promoting its results.

TMNs are the main actors in the production of best climate practice. This is not only the result of the sheer number of municipal networks for climate action,[2] but also of their collective drive to increase the visibility of municipal activities through internet presence, conferences, workshops, seminars, webinars, publications, handbooks, online tools, databases, etc. Here again, political interests are a main reason for the production of best-practice guidance. On the one hand, the premise of TMNs is the exchange of ideas and experiences, and the identification and diffusion of best practice is an essential pillar of their existence in most cases. On the other hand, gathering and disseminating knowledge through best-practice publications, databases and presentations serves to increase the profile of TMNs, their members and their political power (Interview Network-6 2016).

In the empirical case study examined here, the production of best-practice examples among TMNs was considered to be flexible and rather unstructured. These are specifically developed as part of funded projects, or emerge as the result of an ad hoc process involving conversations, conferences and literature, along with personal preference and one's own work through which one comes into contact with relevant people if one needs more information and materials for activities including conferences, publications and databases. It is not necessarily the most innovative or effective examples that are chosen by the organisation, but rather those that entail the widest range of co-benefits. In practice, this includes project information that can be amassed quickly, projects that cover a particular topic, and projects that can be easily communicated. This implies less of an analytical decision-making process based on facts and more a strategic calculation using context-specific criteria.

A similar approach is evident among research institutions and consultancies. The difference is that the production of best practices is usually associated with specific prerequisites and expectations about deliverables. For example, the German environment ministry commissioned the German Institute for Urban Affairs (DIFU) to establish a 'Service and Competence Centre for Municipal Climate Action' (SK:KK). Its aim is to encourage and support climate action in as many municipalities as possible. To this end, a comprehensive online database with practical examples has been established to provide the largest possible range of examples. It enables searches to be conducted according to geographical area, the size of municipalities, and the target audience for the action. The collection of examples is usually linked to dedicated events or research of the funding database of the German Ministry for the Environment. The success of particular actions is confirmed by 'Projektträger Jülich' (PtJ), the designated fund manager of the German Ministry for the Environment (Fieldnotes-150423-Dresden). This example indicates that rather than choosing examples according to carbon-emission reductions, or radically innovative solutions, as the measure of the 'best', the focus instead lies partly on politically motivated choices: for example, practices from a particular geographical location or from a municipality with a declining population, or where particular austerity measures have been imposed, in order to prove that good examples exist in *every* context. The data in Table 5.1 and Figure 5.2 would seem to support this assertion.

The sample comprising 151 best-practice examples from the NKI online database revealed that only seven municipalities host more than one best-practice

Table 5.1 Distribution of good practice examples according to municipal population

Number of best practice examples	*Size of municipality by population*
23	Up to 10,000 inhabitants
36	10,001–50,000 inhabitants
31	50,001–150,000 inhabitants
22	150,001–500,000 inhabitants
10	More than 500,000 inhabitants
29	Districts and regions with 50,000–3,500,000 inhabitants

Source: 'Praxisbeispiele von A-Z' database (www.klimaschutz.de/praxisbeispiele).

Figure 5.2 Spatial distribution of good practice examples.

Source: 'Google Maps' based on 'Praxisbeispiele von A-Z' database (www.klimaschutz.de/praxisbeispiele).

example. Overall, best practices are fairly evenly distributed across population categories (Table 5.1). The plot of best practices by geographical location (Figure 5.2) suggests that the examples may have been chosen to ensure a fairly even spatial distribution.

This raises the question about what criteria are used to determine what makes an example of best practice fit for replication in the studied cases. Several citations and the database examples point towards varied interpretations of selection criteria depending on what a best-practice example is supposed to show. The criteria thus 'depend on the topic and the project', as one interviewee emphasised (Interview Network-5 2016). In the case of inter-city competitions on local climate action, this seems to be different at first glance. The documentation accompanying Germany's federal Ministry for the Environment's 'Klimaaktive Kommune 2009' (climate active municipality) competition, for example, states that:

> Important decision-making criteria were efficiency regarding climate action, transferability to other municipalities and innovativeness regarding specific local circumstances.
>
> (Service- und Kompetenzzentrum: Kommunaler Klimaschutz 2010, p. 12)

Key criteria appear to be climate-action 'effectiveness', the perceived level of innovation, and the potential for replication. A quantitative assessment of the level of innovation and transferability is nearly impossible, and climate-action effectiveness ultimately can only be measured through carbon-emission reductions. The selection procedure, therefore, uses subjective and questionable qualitative selection criteria as opposed to standardised quantitative criteria. Why one example is chosen as 'best practice' and another not has many reasons, as one consultant form a research institute explains:

> It may be linked to the quality of the description or an exciting innovation. It may also be linked to the valuation of progress: which project has evolved most successfully in the last 2–3 years? And that is not necessarily the example that would be likely to win the award again. So, it's not like the Champions League in football where the best team somehow succeeds, but a rather more variable approach.
>
> (Interview Consultant-4 2015)

The development of best practice, therefore, often lacks transparency. According to the cases I studied, neither competitions nor TMNs, research institutions, or consultancies use a standard set of indicators. Unsurprisingly, municipal climate-action representatives often view best-practice examples critically, as this quote makes clear:

> I'm just throwing out there that there are no real criteria, just a number of committees stating 'oh this is best practice!' And this is often linked to the

> presentation, whether it's nice and colourful and it looks impressive and especially if it suggests that thousands of people have been affected by it. That's what really counts, and no one asks about the actual quality or there's just not enough criteria to do so.
>
> (Interview Municipality-19 2017)

Municipal climate-change managers often say that the decision-makers behind best-practice examples use poor and subjective judgement in their choice of what may be considered best practice, as they often lack expertise and context-specific knowledge about municipal challenges (Interview Municipality-18 2017). Thus, they are very sceptical '[...] that truly best practices [in relation to carbon reduction] are being diffused. Instead it is examples that are easiest to communicate' (Interview Municipality-15 2016).

To understand what local knowledge becomes accepted as best practice, and what information is excluded to ensure an example may be considered 'good' or the 'best', it is necessary to take a closer look at content. What types of objects become visible (and invisible) as governable entities through best practice-oriented knowledge production? What types of governance processes are considered appropriate? And how is this information presented?

Of the 368 best-practice examples from three online databases, two publications and one guideline provided by the C40 network, more than one-third are single technical measures, such as replacing a heating system, energy-efficient retrofitting according to Passivehouse standards or installation of LEDs. One-third are examples of long-term mitigation, adaptation or sector-specific strategies (e.g. mobility or tourism). Twenty per cent are measures involving citizens such as educational projects, campaigns and public participation. The examples cover a diverse spectrum of activities. It is noticeable, however, that measures to minimise resource and energy demand by reducing consumption (so-called demand-side management measures) are neglected entirely. Similarly, there are very few examples of new legislative or financial measures.

Reviewing best-practice examples has revealed some insights regarding presentation. Given that climate change management focuses primarily on measurable criteria (such as carbon emissions, cost-benefit calculations and energy savings), it is noteworthy that 55 per cent of best-practice descriptions provide hardly any quantitative data on their performance. Besides, transferability, a point strongly emphasised in funding criteria, was rarely mentioned in the sample examined, with transferability being described in only 20 per cent of cases. The assessment also revealed that 80 per cent of best practices are portrayed entirely as success stories with no background information regarding barriers, challenges or problems throughout the implementation process, despite this information having a high value to municipal representatives (Interview-Municipality-4, -8 2015). Best-practice descriptions – whether on paper or in oral presentations at conferences – do not accurately reflect the problems, challenges and barriers that new projects typically run into. Information about endless hours of work involving phone calls and networking activities to operationalise a project is systematically withheld. Interviewees suggested, for example, that: 'best practice does not know failure' (Interview Municipality-9 2015) and that

> '[…] mediocre and semi-functioning measures are presented as state-of-the-art' because 'every city that experiments with certain actions that other cities have not yet tried and in spending public money for this has always to justify towards their citizens and the political opposition what the money was spent on.'
>
> (Interview Municipality-1 2015)

My investigations made it evident that technologies, methods, processes and procedures that are marketed as best practice are less innovative, radical and transformative than is often portrayed. Best-practice examples produce primarily system-enhancing, as opposed to system-challenging, information (Radaelli 2004; Veselý 2011). One interviewee took the example of the 'Klimaaktive Kommune' competition, for which the federal environment ministry and the DIFU chooses the winners. He observed that in this context 'it is unlikely that a project involving activism [like for example a protest against the car industry], will ever be considered "good practice"' (Interview Consultant-1 2015). This can limit the potential of best practices to contribute to significant transformation of structures, processes or technologies. As one interviewee (Interview Consultant-5 2015) stated:

> I think you have certain standards and best practice examples are chosen in such a way that they come as close as possible to these standards. I do not think that best practices have such an impact that they can significantly influence or shape these standards; they are not strong enough. You always try to come close with your examples to existing standards. […] Therefore, I am not convinced that the impact of best practice exchange is as high as we might desire.

Thus, governing through best practice risks implanting a low level of ambitiousness in governance regimes, as opposed to nourishing creativity and inventiveness (Interviews Consultant-1, Municipality-1 2015; Municipality-15 2016).

Discussion

> The process of production of 'good practices' is at once regulative (influencing cognitive frames, rules and collective representations) and constitutive (producing social relations and identities).
>
> (Vettoretto 2009, p. 1079)

The analysis has shown that identifying, developing and diffusing best-practice examples is a highly political process. 'Best practice' is not a self-evident classification process; practices are always made to appear as the 'best' through the actions of actors who represent various political and strategic interests (Interview Network-1 2015). In short, best practice is the result of social relations and the power dynamics underlying these relations. But two issues remain to be discussed: first, the role of best practices in configuring political rationalities of urban climate governance; second, the transformative potential of best practices in governing urban sustainability transitions.

First, to better grasp the role of best practices in building epistemic authority about what 'good' urban climate-governance means, we need to understand that governance relies on a particular understanding of what constitutes governing practices, as well as their effectiveness. Governance is, thus, dependent on knowledge regarding the object of governance, the place-specific context of governance, the necessary resources for governing, and the interests of the involved actors. According to this interpretation, best practice – as an epistemic technique – creates expectations and defines political standards or intervention processes, producing visibilities that prioritise specific solutions and designs of urban systems while obscuring alternative ones (Voß and Freeman 2016). This raises the following questions: How is governance information produced through best practice? How does best practice, in turn, influence governance practices?

Best-practice examples contribute to the power–knowledge complex in municipal climate action, due to their role in formalising generalisable recommendations for action from individual, local experiences. As a knowledge-producing governmental technology, they appear to significantly influence the decision as to which municipal climate actions and interventions should be deemed reasonable and sensible, and which should not. Instead of objective descriptions, best-practice examples provide selective and powerful interpretations of optimal municipal climate action, as my empirical analysis has already shown. Rarely, it would seem, is best practice a transparent tool that enables governance actors to effectively learn how to successfully implement optimal climate-action procedures. Instead of fostering systematic replication processes to help improve existing governance practices, best practice is used as a politically motivated governmental technology to increase the visibility of particular practices and particular municipalities. Best practices are consequently used to depict supposedly exemplary ways of acting to ensure that specific themes get maximum political attention, or to establish them to overcome argumentative, political or mental barriers, and to make particular governance approaches possible in the first place (Nagorny-Koring 2018).

The preparation, diffusion and use of best-practice examples are accompanied by a discursive process that creates new knowledge and information about both the political problem and the nature of the problem itself (Bulkeley 2006). This implies that neither the choice and coding, nor the oral presentations and publications of best practice represent a neutral act of description. Instead, these processes reflect what is politically considered true/good or wrong/bad within the dominant governance rationality (Dzudzek2016). Language and text need to be understood as performative in this context as they provide the basis for action and creation of social realities. Similar to models and theories, management methods such as best practice not only describe social realities but are also a component of their creation:

> Thus, knowledge production, maintenance, and use are not only epistemic issues; they are simultaneously ontological endeavours. They are acts of making, rather than merely depicting, what is at stake, and, in this sense, they are performative to the core.
>
> (Passoth and Rowland 2006, p. 38)

Best-climate practices influence municipal climate action instead of simply depicting which measures have been implemented successfully. Assigning the best-practice label per se means assigning power and influence to a practice (Veselý 2011). This can imply that the chosen practices and examples only appear remarkable following their designation as 'best'. By repeating the identification of best practice in databases, journals, seminars and conferences, they are legitimised and verified (Overman and Boyd 1994). Discussing and diffusing best-practice cases leads to their normalisation and reproduction, which enhances the performative effect (Bulkeley 2006; Moore 2013).

Best practices are, hence, not passive instruments waiting for application in a toolbox. They are compromised packages of knowledge produced through, as well as within, social networks comprising scientific, political, advisory, administrative and economic actors. These actors create a political epistemic culture, which in turn creates particular realities of governance (Voß 2016). Knowledge regarding climate governance is materialised and stabilised through description and coding of best-practice examples; they are therefore an important aspect of the climate-governance knowledge-production infrastructure (Simons 2016). According to Voß (2016, p. 129), governmental technologies, such as best practice, do not engender aspects of a naturally persisting political reality; they 'programme' these political realities. Thus, best practices co-create, shape and structure the rationales underpinning the management of urban low-carbon transitions by rendering the diffuse phenomenon of climate change as stable, mobile, combinable and comparable information. Consequently, best practice establishes and stabilises certain representations of political reality by providing knowledge about 'successful' political practices. In doing so, they act as a '[...] means through which discourses and visions concerning the future of cities are rendered practical, and governable' (Bulkeley and Castán Broto 2013, p. 367), ultimately contributing to building epistemic authority about what is 'good' urban climate governance.

Second, and in line with the previous discussion, the normative question inevitably arises as to how best practice exclusively reproduces political realities and thereby stabilises systems, or whether it can also contribute to changing these rationalities and thus to transforming social reality. Therefore, I now go on to consider if best practice is instrumental in cultivating governing rationalities in support of far-sighted system transitions, or alternatively, if it is rather a barrier to urban sustainability transitions by cultivating fragmentation and short-sightedness.

Although it is unclear how 'transformative' best-practice cases are, the expectations placed on them, and hence the demand for examples, are generally high. This is evident in the statement by a TMN representative who argues that best practice '[…] is something that can help drive down the costs and help some cities to skip the whole kind of history of development that European cities have passed through' (Interview Network-6 2016). Here, governing through best practice quickly becomes a 'stage-managed consensus' (MacLeod 2011, p. 2632), representing the urban sustainability transition as manageable with the help of working good-practice examples.

However, my analysis of best practice suggests that best-practice examples may rarely contribute to structural change. Neither the most innovative nor radical ideas, nor measures with the greatest carbon emission-reduction potential are routinely selected and disseminated as best practices. Rather, they tend to be approaches that can be readily communicated, and are easy to implement, the most data-rich, or the most cost-effective. This can lead to innovations being thwarted and urban visions that cement the social status quo being promoted (Veselý 2011; Moore 2013). This is in line with some of the climate-change managers' impressions, who told me that best-practice examples are often repetitive and only a few actual innovations are communicated, which means their transformative potential is low (Interviews Municipality-16 2016; Municipality-18, -19 2017). In effect, the politics and knowledge-configuring practices underpinning the identification of best practice appear to support techno-managerial, fragmented and short-term project-oriented forms of governing, which do not challenge the logic and routines of the existing socio-political system and its constitutive power relations (Bruno 2009; Moore 2013).

Conclusion

This chapter seeks to answer the research question: How is knowledge politics performed through the promotion of 'best practice' as a particular governing technique intended to develop system-transforming ways of knowing the urban fabric? The aim was to investigate the complex politico-epistemic practices through which instances of local urban experimentation are identified and abstracted into 'best practice for urban sustainability transition'. I, thus, concentrated on revealing the hidden production processes of best practices to uncover how epistemic technologies are currently involved in co-defining the urban fabric. To do this, I showed empirically how authoritative knowledge about urban climate-change governance is politically constituted through best practice, using the example of the German funding programme 'Masterplan 100% climate protection'. Thus, on the one hand, this chapter contributes to the emerging research into the knowledge politics of urban sustainability transitions as outlined in this book (see Chapter 1). On the other hand, the governmentality perspective being applied in this chapter contributes to revealing the underlying power–knowledge complex of best practice and helps to better understand how governing climate change operates on the ground.

I opened this chapter by introducing best practice as a popular and powerful knowledge-configuring device that lacks critical scrutiny but seems to actively co-define the nature of urban reality in line with the political-strategic interests of its 'producers' and 'users'. The introduction was followed by an outline of governmentality as the theoretical lens I applied to illustrate the constitutive role of knowledge in urban climate governance, to explore the role of epistemic devices in the production and enactment of authoritative knowledge, and to critically reflect on the suitability of governing by best practice to accelerate urban sustainability transitions. After making my methodology transparent, I dealt in the analysis part with the question as to how best practice for municipal climate action is created out of everyday governance. As a result, the chapter's main findings are that 'best practice' is not a transparent epistemic tool whereby governance actors may effectively learn how to optimally orchestrate transition experiments. Rather, best practice is a political technology that urban actors use to render particular cities, governing solutions, strategies and certain political agendas visible and attractive.

In the discussion section, I revealed, based on the existing literature and my own empirical findings, two important aspects of the performance of knowledge politics through best practice. First, concerning the role of best practices in configuring political rationalities of urban climate governance, it became clear that best-practice examples contribute to defining political standards and intervention processes, thereby producing visibilities that prioritise specific solutions and designs of urban systems while obscuring alternative ones. Best practices, thus, co-shape and co-structure the governing rationales of urban low-carbon transitions by creating and disseminating certain representations of political reality, and by providing knowledge about the 'successful' political practices. Second, concerning the transformative potential of best practice in governing urban sustainability transitions, the discussion of the findings made clear that governing by best practice has to be seen rather as a barrier to urban sustainability transitions, because it cultivates fragmentation and short-sightedness and, thus, only rarely contributes to structural change. The politics and knowledge-configuring practices underpinning the identification of best practice appear to support techno-managerial and project-oriented forms of governing that do not challenge the existing socio-political system and its constitutive power relations.

Finally, the discussion led me to conclude that uncovering how best practices performatively shape political rationalities opens up a new understanding of the stickiness of the status quo. However, one should not underestimate the importance of storytelling through best practices as an epistemic device, as a means of repeating and naming reality. Emphasising the effect of language and narration highlights the subjects' active and constitutive ability to change themselves and society (Livesey *et al.* 2009). The challenge in identifying, selecting, processing and disseminating best practice is then less to describe what current solutions are and how they work, but more to report how change can succeed, which hurdles to overcome, and what an alternative future could look like. But currently, best practice often still functions as a political sedative rather than an engine of transition.

Acknowledgements

I gratefully acknowledge funding from the EIT Climate-KIC for conducting this research related to my PhD project 'Practice Regime Municipal Climate Governance: Governing by Best Practices'. Special thanks go to Colin Nolden for the translation of this book chapter and Elke Alban for the cartography. Moreover, I would like to thank the editors of this book – Jens Stissing Jensen, Matthew Cashmore and Philipp Späth – for organising, and inviting me to, the workshop on 'Epistemic Politics of Urban Sustainability Transitions', and giving intense and constructive feedback on this chapter.

Notes

1 Namely: C40, Climate Alliance, Metropolis, Energy Cities, Local Governments for Sustainability (ICLEI), and the Covenant of Mayors.
2 For example, Energy-Cities, ICLEI, C40, Climate Alliance, Covenant of Mayors, R20, The Climate Group, Resilient Cities as well as working groups in Eurocities, Metropolis, Alliance in the Alps, Union of the Baltic Cities.

Interviews

Interview-Consultant 1, 2015. Personal interview on 17 March with senior project manager, adelphi research. Transcript in possession of author.
Interview-Consultant 4, 2015. Personal interview on 17 June with project lead, deEnet (Kompetenznetzwerk dezentrale Energietechnologien e.V.). Transcript in possession of author.
Interview-Consultant 5, 2015. Personal interview on 26 November with project manager, adelphi research. Transcript in possession of author.
Interview-Municipality 1, 2015. Personal interview on 11 June with climate protection officer. Transcript in possession of author.
Interview-Municipality 4, 2015. Personal interview on 18 August with Masterplan implementation manager. Transcript in possession of author.
Interview-Municipality 8, 2015. Personal interview on 18 November with head of climate protection unit. Transcript in possession of author.
Interview-Municipality 9, 2015. Personal interview on 18 November with head of regional climate protection unit. Transcript in possession of author.
Interview-Municipality 15, 2016. Personal interview on 7 June with Climate-KIC project manager. Transcript in possession of author.
Interview-Municipality 16, 2016. Personal interview on 10 June with climate protection officer. Transcript in possession of author.
Interview-Municipality 18, 2017. Phone interview on 27 January with climate manager. Transcript in possession of author.

Interview-Municipality 19, 2017. Phone interview on 2 February with head of energy department. Transcript in possession of author.
Interview-Network 1, 2015. Personal interview on 12 October with project manager policy transfer. Transcript in possession of author.
Interview-Network 5, 2016. Personal interview on 1 March with head of urban research. Transcript in possession of author.
Interview-Network 6, 2016. Skype interview on 5 July with regions project coordinator. Transcript in possession of author.

Fieldnotes

Fieldnotes-150423-Dresden, 2015. Notes taken during attendance of Climate Alliance annual international conference on 15 April in Dresden. Notes in possession of author.
Fieldnotes-160601-Berlin, 2016. Notes taken during attendance of Metropolitan Solutions fair on 1 June in Berlin. Notes in possession of author.

References

Bendixsen, S. and Guchteneire, P., 2003. Best practices in immigration services planning. *Journal of Policy Analysis and Management*, 22 (4), 677–682.
Benz, A., Kemmerzell, J., Knodt, M. and Tews, A., 2015. The trans-local dimension of local climate policy. Sustaining and transforming local knowledge orders through trans-local action in three German cities. *Urban Research & Practice*, 8 (3), 319–335.
Boulanger, S. and Nagorny, N., 2018. Replication vs mentoring: accelerating the spread of good practices for the low-carbon transition. *International Journal of Sustainable Development & Planning*, 13 (2), 316–328.
Brand, P., 2007. Green subjection: the politics of neoliberal urban environmental management. *International Journal of Urban and Regional Research*, 31 (3), 616–632.
Bruno, I., 2009. The 'indefinite discipline' of competitiveness benchmarking as a neoliberal technology of government. *Minerva*, 47 (3), 261–280.
Bulkeley, H., 2006. Urban sustainability: learning from best practice? *Environment and Planning* A, 38 (6), 1029–1044.
Bulkeley, H. and Castán Broto, V., 2013. Government by experiment? Global cities and the governing of climate change. *Transactions of the Institute of British Geographers*, 38 (3), 361–375.
Bundesministerium für Umwelt, Naturschutz, Bau und Reaktorsicherheit (BMUB), 2015. *Merkblatt Förderung von Masterplan-Kommunen: Hinweise zur Antragstellung*. Available at: www.klimaschutz.de/sites/default/files/merkblatt_masterplan-richtlinie.pdf [accessed 6 October 2017].
Bundesministerium für Umwelt, Naturschutz, Bau und Reaktorsicherheit (BMUB) ed., 2017. *Nationale Klimaschutzinitiative: Zahlen und Fakten*. Available at: www.klimaschutz.de/zahlen-und-fakten [accessed 6 October 2017].
Demeritt, D. and Nobert, S., 2014. Models of best practice in flood risk communication and management. *Environmental Hazards*, 13 (4), 313–328.
Dzudzek, I., 2016. *Kreativpolitik: Über die Machteffekte einer neuen Regierungsform des Städtischen*. Bielefeld: Transcript.
EU, 2016. *Urban agenda for the EU: Pact of Amsterdam*. Available at: http://ec.europa.eu/regional_policy/sources/policy/themes/urban-development/agenda/pact-of-amsterdam.pdf [accessed 23 January 2018].

Foucault, M., 1995. *Discipline and punish: The birth of the prison*. New York: Vintage.

Liem, A.D. and McInerney, D.M., eds, 2008. *Teaching and learning: International best practice*. Charlotte, NC: Information Age Publishing.

Lindblom, C.E., 1959. The science of 'muddling through'. *Public Administration Review*, 19 (2), 79–88.

Livesey, S.M., Hartman, C.L., Stafford, E.R. and Shearer, M., 2009. Performing sustainable development through eco-collaboration: The Ricelands Habitat Partnership. *Journal of Business Communication*, 46 (4), 423–454.

MacLeod, G., 2011. Urban politics reconsidered: growth machine to post-democratic city? *Urban Studies*, 48 (12), 2629–2660.

Macmillen, J. and Stead, D., 2014. Learning heuristic or political rhetoric? Sustainable mobility and the functions of 'best practice'. *Transport Policy*, 35, 79–87.

Montero, S., 2017. Worlding Bogotá's Ciclovía: from urban experiment to international 'best practice'. *Latin American Perspectives*, 44 (2), 111–131.

Moore, S., 2013. What's wrong with best practice? Questioning the typification of new urbanism. *Urban Studies*, 50 (11), 2371–2387.

Nagorny-Koring, N., 2018. Leading the way with examples and ideas? Governing climate change in German Municipalities through best practices. *Journal of Environmental Policy & Planning*, 5 April. doi.org/10.1080/1523908X.2018.1461083.

Nagorny-Koring, N. and Nochta, T., 2017. Managing urban transitions in theory and practice: the case of the pioneer cities and transition cities projects. *Journal of Cleaner Production*, 175, 60–69.

Overman, S. and Boyd, K., 1994. Best practice research and postbureaucratic reform. *Journal of Public Administration Research and Theory*, 4 (1), 67–84.

Passoth, J.-H. and Rowland, N., 2006. Modeling the state: an actor-network approach. In J.-P. Voß, D. Bauknecht and R. Kemp, eds. *Reflexive governance for sustainable development*. Cheltenham: Edward Elgar, pp. 37–61.

Patton, M.Q., 2002. *Qualitative research and evaluation methods*. Thousand Oaks, CA: Sage.

Radaelli, C., 2004. The diffusion of regulatory impact analysis – best practice or lesson-drawing? *European Journal of Political Research*, 43 (5), 723–747.

Rose, N., O'Malley P. and Valverde M., 2006. Governmentality. *Annual Review of Law and Social Science*, 2 (1), 83–104.

Schechtman, J., 2015. Best practices for cost-efficient coastal climate change adaptation after damaging storms. *Journal of the American Planning Association*, 80 (4), 336–337.

Schreurs, M., 2008. From the bottom up: local and subnational climate change politics. *The Journal of Environment & Development*, 17 (4), 343–355.

Service- und Kompetenzzentrum: Kommunaler Klimaschutz, 2010. *Kommunaler Klimaschutz 2009: Wettbewerb > Die Preisträger und ihre Projekte*, Deutsches Institut für Urbanistik. Available at: https://difu.de/publikationen/2010/kommunaler-klimaschutz-2009.html [accessed 16 January 2018].

Simons, A., 2016. Fact-making in permit markets: document networks as infrastructures of emissions trading. In J.-P. Voß and R. Freeman, eds. *Knowing governance*. London: Palgrave Macmillan, pp. 177–192.

Stead, D., 2012. Best practices and policy transfer in spatial planning. *Planning Practice and Research*, 27 (1), 103–116.

Stripple, J. and Bulkeley, H., 2014. Introduction: on governmentality and climate change. In J. Stripple and H. Bulkeley, eds. *Governing the climate: New approaches to rationality, power and politics*. New York: Cambridge University Press, pp. 1–23.

Tedesco, C., 2010. EU and urban regeneration 'good practices' exchange: from download to upload Europeanization? In A. Hamedinger and A. Wolffhardt, eds. *The Europeanisation of cities: Policies, urban change, and urban networks*. Amsterdam: Techne Press, pp. 183–195.

Veselý, A., 2011. Theory and methodology of best practice research: a critical review of the current state. *Central European Journal of Public Policy*, 5 (2), 98–117.

Vettoretto, L., 2009. A preliminary critique of the best and good practices approach in European spatial planning and policy-making. *European Planning Studies*, 7 (17), 1067–1083.

Voß, J.-P., 2016. Realizing instruments: performativity in emissions trading and citizen Panels. In J.-P. Voß and R. Freeman, eds. *Knowing governance*. London: Palgrave Macmillan, pp. 127–153.

Voß, J.-P. and Freeman, R., 2016. Introduction: knowing governance. In J.-P. Voß and R. Freeman, eds. *Knowing governance*. London: Palgrave Macmillan, pp. 1–33.

Walters, W., 2012. *Governmentality: Critical encounters*. London: Routledge.

Ziesing, H.-J., 2010. *Masterplan 100% Klimaschutz: Auf dem Weg zur Null-Emissions-Kommune*. Available at: www.ptj.de/lw_resource/datapool/_items/item_6255/hintergrundpapier_masterplan.pdf [accessed 16 January 2018].

6 Competing knowledge assemblages in Danish heat governance

Jens Stissing Jensen[1] *and Peter Karnøe*[2]

Introduction

Contemporary imaginaries of future low-carbon energy systems are often populated by wind turbines and solar panels. Decarbonisation of energy systems based on such production technologies presupposes the distribution and use of electrical energy and that these systems are made compatible with the weather-dependent production patterns of renewable technologies. Over the last decade energy governance has been increasingly concerned with how to conceptualise distribution infrastructure and end-use domains as smart building-blocks of the energy system that can be made compatible with intermittent power production.

So-called smart power grids that can be controlled and optimised through a combination of modern ICT technology and real-time price signals have been heavily promoted as an important component of low-carbon energy systems (JRC Science Hub 2017). In Denmark alternative imaginaries of the building-blocks of future low-carbon energy have also been cultivated. Energy-system models have indicated that heat provision (and cooling) of buildings should be considered a 'smart' component in adapting energy systems to fluctuating low-carbon power production (Lund *et al.* 2014). By converting electricity into hot water that can be stored for days, weeks or months, it has been argued that district heating infrastructures can provide affordable flexibility, and in large volumes.

Much discourse on how to develop distribution infrastructures and end-use domains is conducted within idealised laboratory-like imaginaries of technical energy-system models. In some cases, the smart building-blocks cultivated under such laboratory assumption have been tested in pilot experiments (EcoGrid eu 2016). However, it is rarely discussed how novel and innovative ideas about the role of distribution infrastructures and end-use domains in low-carbon energy systems can be integrated into the governance activities through which energy systems are concretely reproduced and transformed on an everyday basis.

In response to the relative neglect of the real-world everyday governance of low-carbon energy transitions, the purpose of this chapter is to analyse how political rationalities pertaining to energy governance are constructed and controlled. We also discuss how such rationalities may be transformed in response to new societal objectives, such as achieving carbon neutrality. More specifically, we explore how political rationalities pertaining to one specific building block of the Danish energy system – heat provision – have been developed, shaped and contested over the past 40 years.

In doing this we pay specific attention to how political rationalities and the resulting governance activities are predicated on the ways in which heating is rendered knowable. This is because system governance is fundamentally a knowledge-dependent activity. Consider, for example, the complex types of knowledges that are needed to identify, value and compare different investment options in heating, given their long technical lifetime, their many technical interdependencies with the broader energy system and their relations to political objectives. In the following analyses we argue that the knowledge-configuring devices and procedures that actors have at their disposal in qualifying such mundane decision situations are an essential political ingredient of heat governance because such devices and procedures constitute a means to effectively control and shape the rationalities that inform planning and investment practices.

Our chapter shows how political rationalities in Danish heat governance are intrinsically predicated on particular ways of knowing. We furthermore illustrate how knowing necessarily depends on a network of contentious analytical assumptions, and that knowing only becomes authoritative as long as this network is effectively policed by expertise and political interests. Finally, we propose that the ability to strategically question and rearrange analytical assumptions involved in knowing is a central ingredient in reconfiguring political rationalities and resulting governance activities.

More specifically our analysis illustrates how dominant knowledge-producing activities controlled by the national governmental administration promote a market-imitating and efficiency oriented political rationality. Recently, city regions in Denmark have attempted to politicise this rationality for neglecting concerns over carbon emissions. This has involved attempts to reconstruct knowledge production at the city scale in support of coordinated, low-carbon transitions. So far attempts to politicise the dominant political rationality through alternative ways of knowing have however not succeeded to fundamentally reorganise actual planning and investment activities. The interests and expertise invested in policing the dominant analytical assumptions thus succeeds to marginalise alternative ways of knowing. We conclude that future low-carbon transition of Danish heat provision is likely to be influenced by the eventual outcome of the low-carbon knowledge controversies between the national governmental administration and local city administrations.

The knowledge-politics of contemporary governance

Our focus on knowledge production as a central means to construct, control and shape political rationalities in relation to contemporary energy-system governance is inspired by a recent knowledge turn in the governance literature (Voß and Freeman 2016). This turn sees knowledge production as an activity that is less concerned with passively mirroring social reality and more concerned with rendering reality visible in strategically controlled ways, by closing down and purifying the inherent complexities and ambiguities of real-world situations. Mitchell (2008), for example, argues that economic expertise and economic calculation should not always be seen as passive representations of pre-existing realities, but rather stipulates that such calculations are often strategically constructed to render new realities conceivable. He, for example, illustrates how Edison – the inventor of urban electricity systems – made extensive use of economic calculations in order to render such systems comprehensible to investors. In this case, economic calculations were less about accurately representing a system and more about making a novel network intelligible to a particular actor-grouping:

> Since these calculations were helping to bring into being the world they calculated, success did not necessarily depend on having the most accurate figures. What mattered more was whether the calculations enabled the network to be conceived and built.
>
> (Mitchell 2008, p. 1117)

Recognition that knowledge production involves purifications that are designed to make complex and ambiguous state-of-affairs appear as transparent and manageable objects has called attention to the devices and procedures that are involved in producing these purifications. Miller and Rose (1990) note that:

> Knowing an object in such a way that it can be governed is more than a purely speculative activity: It requires the invention of procedures of notation, ways of collecting and presenting statistics, the transportation of these to centres were calculations and judgements can be made and so forth.
>
> (Miller and Rose 1990, p. 5)

Along similar lines of reasoning, Bueger (2015) argues that analysis of contemporary governance requires attention to the concrete micro-physics of knowledge production. His analysis accordingly focuses on how knowledge-configuring devices and procedures are arranged in order to construct and delineate new governable universals 'by drawing different particulars together, including data, facts, and claims' (Bueger 2015, p. 6). Such attention to the micro-physics of knowledge production hints at the fundamentally political character of knowing: bringing together a specific set of particulars (while ignoring others) in knowledge-configuring devices has a decisive impact on which type of concerns, actors and experiences are rendered visible in, and germane to, governance. According to the knowledge turn, a salient aspect of contemporary political change has to do with re-coding knowledge-configuring devices in order to selectively redefine how social reality is rendered visible, and which concerns, actors and experiences are made to count.

With these reflections as point of departure, the remaining parts of this chapter investigate how heat provision in Denmark has been rendered governable over the past 40 years under different, and sometimes competing, knowledge assemblages. We define a knowledge assemblage as a relatively well-structured network of knowledge-producing devices and procedures that is instrumental in rendering open-ended and opaque states of affairs conceivable as well-defined and actionable decision situations where options for action can be identified and prioritised.

Our analysis of the relation between knowing and political rationalities pertaining to heat provision in Denmark is informed by three main empirical datasets. The fist dataset consists of a series of policy documents issued since 2001 focusing on how to modernise heat governance and an interview with an analyst from the Danish Energy Agency. The policy documents reflect the introduction a set of calculative devices promoted to frame and assess heat provision as market. The second dataset consists of policy documents that aim to configure knowledge in response to politically articulated carbon concerns by national and urban governance arrangements. The final dataset consists of six interviews with heating utilities and actors involved in urban heat-governance initiatives. These interviews explore how knowledges are applied to inform and qualify actual planning activities and investment decisions.

A knowledge-politics biography of heat provision governance in Denmark

The initial development of networked heating infrastructure in Denmark dates back to the 1920s, when the first district-heating systems were constructed to use waste heat from urban power plants to heat nearby buildings. It was not until the adoption of a national heat law in 1979 that heating was articulated in a coherent manner as a matter of national policy concern. This law was adopted as a response to society's dependency on oil as a heating fuel, which had become a salient liability for the national economy as international oil prices rocketed due to a series of international oil crises (Danish Energy Agency 2016).

The heat law of 1979 was instrumental in initiating the development of governance arrangements that, over the following decade, served to orchestrate large-scale investments in networked heating infrastructures based on natural gas and district heating, with the objective of decreasing national dependency on imported oil. A prerequisite for the planning of investments activities was the development of a knowledge assemblage capable of rendering heating visible as a coherent infrastructural challenge. To this end, highly structured knowledge-production practices were developed in which municipalities were responsible for mapping local heat demands and the heating technologies used. Based on these inputs, regional authorities developed heat plans in collaboration with municipalities and heating utilities. These regional plans had to comply with planning prerequisites issued by the Danish Ministry of Energy (Danish Energy Agency 2017).

The municipalities then translated these regional heat plans into local municipal plans which constituted a legally binding basis for the approval of individual investments in heating infrastructures such as heat-production facilities and distribution grids for gas and district heating. The regional heat plans contained guidelines for: (a) which areas to be heated by district heating, natural gas and individual heating solutions; (b) where to locate heat-production facilities and heating infrastructures and; and, (c) where to locate industries with high production or consumptions of heat. All in all, heat planning was underpinned by a comprehensive and centrally orchestrated planning infrastructure that included coordinated knowledge-configuring activities among local, regional and national planning bureaucracies (Danish Energy Agency and Local Government Denmark 2010). The magnitude of these planning efforts is reflected in the amount of working hours that were invested in heat planning. Eighty employees were employed on heat planning in the National Energy Agency alone, while each region had two, and each municipality had one full-time employee working with heat planning. From 1981 to 1987 the planning activities were supported by monthly meetings in 16 coordination groups that comprised representatives from the National Energy Agency (a part of the Ministry of Energy), regional authorities, municipal authorities and heating utilities (Danish Energy Agency and Local Government Denmark 2010).

Danish heat governance in 1980s thus illustrates how elaborate and costly knowledge-producing activities were needed in order to bring heat provision into being as an operable infrastructural governance object.

The construction of a market imitating knowledge assemblage for heat-provision investments

During the 1980s spatial heat plans constituted a centrally controlled knowledge assemblage. This assemblage made heating intelligible as an infrastructural coordination challenge where fragmentation in nationwide investments needed to be effectively counteracted. As the overall national layout of heat zones and the associated infrastructural developments were perceived to be completed towards the end of the 1980s, this infrastructural intelligibility was gradually rendered redundant. This was reflected in a series of reforms that weakened the role of spatial heat plans in governance. The first step in this development was a reform from 1990, which authorised municipalities to approve individual investment decisions in heat provision *without reference to regional and national heat plans*. Additionally, municipalities were no longer obliged to develop local heat plans and, in case where they did so, these plans were no longer legally binding (Danish Energy Agency and Local Government Denmark 2010).

This weakening of the infrastructural intelligibility did not imply, however, that heating disappeared as a coherently delineated governance object. Heat governance was transformed and redeveloped around a very different knowledge assemblage that aimed to render heating utilities into market-imitating actors responsible for identifying 'optimal' investments on an independent basis. The challenge in making this market-imitating rationality operational was that heating utilities did not operate as individualised profit-seeking actors in a competitive marketplace. First, the heating utilities enjoyed a natural monopoly, because only a single network of heating infrastructure was allowed to be developed in a given area. Regulation made it mandatory in many cases for properties to be connected to collective heating infrastructures in areas where such infrastructures were available. Furthermore, the regulations stipulated that utilities were not allowed to make a profit. Any profit should be reinvested to reduce the prices for end-users the following year. Heating utilities were also owned by end-users and by municipalities (Danish Energy Agency and Danish Competition Authority 2004).

The ambition to transform these utilities into market-like investors was reflected in, amongst other things, the Project Consolidation Act of 2004 which established a set of market-imitating calculative procedures as central tools in the governing of investment decisions (Danish District Heating Association 2015a). These mandatory calculative procedures had a dual objective. First, utilities were required to calculate the consequences of new investment options for the energy bills of their end-users. Second, the utilities were required to demonstrate that investments constituted an optimal allocation of resources for society as a whole (Danish District Heating Association 2015b). To this end, the actual economic price signals experienced by utility companies would not work, as these prices typically did not reflect the full societal costs. Some energy resources were heavily taxed, while other were not taxed at all. Furthermore, externalities such as air pollution were not included in the market prices. To avoid societally suboptimal investments, the new act stipulated that only investment options demonstrated to be favourable when economised in so-called *socio-economic costs* – i.e. excluding taxes but including major externalities – could be approved by the municipality, who constituted the planning authority.

The requirements for market-imitating calculations presupposed a planning future in which price signals were reasonably predictable. It was a central responsibility of the Danish Energy Agency to construct and maintain a catalogue of these price signals (Danish Energy Agency 2017). The specific knowledge-configuring activities involved in the construction of these price signals illustrates how price signals are seldom a neutral reflection of societal resource scarcity, but more often are complex planning objects influenced by potentially contentious analytical assumptions.

First, in order to construct price signals covering the technical lifetime of investments in heat provision (typically more than 25 years), a qualified guess needed to be made regarding the future development of the national and global energy system, including future developments in the demand and supply of energy and fuels. The Danish Energy Agency chose two specific energy-system scenarios for this purpose. One of these was a baseline scenario for the development of the global energy system called 'new policy scenario' (International Energy Agency 2016, 2017), which was issued by the International Energy Agency. The other was the Danish Energy Agency's own baseline scenario for the likely development of the Danish energy system (Danish Energy Agency 2018).

A key principle of these scenarios was to remain analytically neutral. In practice, this principle promoted status quo-biased analytical assumption, which in turn suggested that the status quo would persist in regard to the energy system's dependency on fossil fuels. While the 'new policy scenario' included some national policies to reduce future carbon emissions, the IEA concluded that these policies would not be 'nearly enough to limit global warming to less than 2°C' (International Energy Agency 2016, p. 2). As the scenario was promoted by IEA as their most central scenario, some actors warned that it could become a self-fulfilling high-carbon prophesy, if investors would regard it a forecast, and adapt their investments accordingly (The Climate Institute 2017; Euractive 2018). While new policy scenarios were based on 'new policy' assumptions, the Danish energy-system scenario was predicated on analytical assumptions that 'policy' (including energy taxes and subsidies) would remain entirely unchanged in the future, irrespective of whether 'current policy' complied with decarbonisation objectives. A consequence of this assumption was that Danish electricity production from wind-turbines – a principal component of efforts to decarbonise the Danish energy system – was anticipated to decrease from 2020 onwards, as no future political agreement about *a continuation* of the feed-in tariffs for wind turbines could be included in the scenario (Danish Energy Agency 2018, p. 11). The 'frozen policy' scenario also assumed that energy taxes and subsidies would remain unchanged in the future (Danish Energy Agency 2018, p. 11), though taxes and subsidies were heavily criticised for hindering socio-economically favourable low-carbon investments. While the Energy Agency publicly proclaimed their baseline scenario to reflect 'the most likely future', conversations with an analyst within the Agency indicated that it was perceived as an 'analytical yardstick' for future policy-making rather than as a prediction of 'the most likely future', as reflected in the interview extract below:

INTERVIEWER: How realistic is the scenario?

INTERVIEWEE: (…) Well, the main idea with frozen policy is that the politicians stop developing new policies, and the system is left to itself. And then what happens? And it is completely unrealistic to imagine that politicians stops making new politics. That is sort of what they are employed to do. It is therefore not a realistic scenario. (…) the baseline scenario is a yardstick you can use to assess new policies.

The energy-system scenarios employed to construct longsighted socio-economic price signals accordingly constituted high-carbon futures developed on the basis of status quo-biased and carbon-friendly analytical assumption. The price signals thus forced heating utilities to plan and economise for an energy system future that did not reflect national or international de-carbonisation ambitions and ignore price signals constructed on alternative (low-carbon) energy-system futures. This high-carbon future, for example, made biomass appear as a relatively attractive fuel, while more ambitious low-carbon scenarios tended to predict an increased global demand – and price – for biomass.

Second, the socio-economic price signals were influenced by the analytical delineation of relevant socio-economic effects. The socio-economic price signals were thus only adjusted for taxes and subsidies levied in Denmark and for externalities occurring on Danish territory. This implied that the negative health effects of, for example, particle pollution from Danish power plants would cease to count as a socio-economic externality once the pollution entered neighbouring territories like Sweden or Germany. A practical effect of this national delineation was that economic prices on imported fuels were not being adjusted for externalities related to their production abroad, such as negative health effects on local workers or local environmental degradation. The socio-economic prices thus systematically ignored costs occurring outside Denmark, though international reports provided increasingly strong documentation of significant subsidies and externalities pertaining to worldwide fuel production (CarbonBrief 2017). The invisibility of such costs systematically disadvantaged for example wind power and solar power in comparison to production technologies that used imported fuels.

Finally, socio-economic price signals were also influenced by the value assigned to externalities, including CO_2 emission and pollution. The valuation of externalities was influences by how they were categorised. CO_2 emissions were for example categorised by the Danish ministry of Finance as a 'policy cost' – i.e. as a cost-imposed policy objective to reduce CO_2 emissions, rather than a cost related to real-world damages from climate change (Danish Energy Agency 2017, p. 22). Following this categorisation, the socio-economic price on CO_2 emissions had to reflect the added cost of realising CO_2 reduction targets. Since the EU had constructed a quota market as a mechanism for reducing CO_2 emissions, this quota price (approximately US$7) was used as the socio-economic cost of CO_2 emissions (Danish Energy Agency 2017, p. 22). The CO_2 quota was also significantly lower than the level recommended by leading international economists in order to keep global warming below 2°C (Stiglitz *et al.* 2017). Also, it was far too low to have a decisive effect on the socio-economic valuation of investment options.

In conclusion, the construction of socio-economic price signals was not the outcome of neutral adjustments of 'distorted' business-economic price signals. Rather, it entailed the prioritisation of specific status quo-based energy-system futures and a delineation of socio-economic effects that rendered cost and externalities located outside Denmark invisible. It also involved a valuation of externalities that made the cost of CO_2 economically insignificant.

In addition to the construction of socio-economic price signals, the market-imitating knowledge assemblage also included devices designed to calculate whether heating utilities succeeded in becoming more 'market efficient'. A central device in this regard was the introduction of a benchmark designed to measure the cost difference between utilities (Danish Energy Agency and Danish Competition Authority 2004). All but the least expensive utilities and infrastructure operators were assigned an efficiency-improvement potential equal to their costs minus that of the most efficient providers. By aggregating the individual efficiency-improvement potentials for all utilities and infrastructure operators, the total market inefficiency for the heating sector emerged as a new knowledge object. The construction of 'market inefficiency' as a quantifiable knowledge object enabled 'the inventive structures' of the heating sector to emerge as a central political-regulatory concern (COWI 2009). It was suggested in governmental reports that the calculated efficiency potentials of the sector could be harvested through additional market-imitating regulations, for example by establishing an individual price ceiling on heating prices issued by each heating utility, in order to force the most expensive utilities to become more efficient (The inter-ministerial working group concerning work regarding comparison of efficiency in the district heating sector 2015).

Our discussion has illustrated how the infrastructural intelligibility cultivated in the 1980s was actively re-coded from the 1990s onwards, with the aim of rendering heating utilities into individualised pseudo market investors. This re-coding converged on a market-imitating knowledge assemblage that linked together objects and concerns such as 'incentive structures', 'efficiency potentials', 'socio-economic price signal', 'baseline futures' and 'efficient allocation of resources'. In the context of this knowledge assemblage, it became increasingly difficult to formulate political objectives for the development of the heating sector other than further market-oriented optimisation. The role of investments vis-à-vis the energy system as a whole was also reframed. Whereas heat investments were conceived as orchestrated activities that should contribute to actively shape the energy system in the 1980s, the market-imitating knowledge assemblage made the energy system intelligible as an external market context for individualised investments. Investment decisions were no longer perceived to actively contribute to a coordinated transformation of the energy system. This market context was made calculable in the guise of price signals that were derived from status quo-based energy-system scenarios that fundamentally failed to meet international CO_2 reduction targets. A path-dependent knowledge assemblage had been constructed with 'market efficiency' as its defining visibility. This assemblage co-produced a series of invisibilities including, amongst other things, democratic control, local engagement and climate impact. Such concerns were rendered invisible both as knowledge objects and as legitimate and governable political concerns.

Urban knowledge assemblages for low-carbon heat provision

While the market-imitating knowledge assemblage described in the previous subsection became increasingly hegemonic, this assemblage did not develop in a vacuum. From 2008 onwards a growing national and international concern for climate change emerged. This concern became particularly influential as a new Energy Act was endorsed by a large majority of parties in the national parliament in 2012. This act stipulated that the Danish energy system, including heating and transport, should become carbon neutral by 2050. An interim ambition was to render the heating and power sectors carbon neutral by 2035. Thus, while CO_2 emissions were easily outweighed when valued using the socio-economic price signals issued by the Energy Agency, the new Energy Act reframed carbon emissions into a politically salient concern that should in principle be addressed regardless of costs.

This political revaluation of carbon emissions could not be ignored, despite its contradictions with the established knowledge assemblage. The carbon targets were not translated into alternative business and socio-economic price signals. Rather, a series of *technical* energy-system futures were constructed (Danish Energy Agency 2014). These future scenarios suggested that a carbon-neutral energy system could be achieved either by (a) drastically increasing the use of (imported) biomass in traditional thermal heat and power plants or (b) by increased reliance on wind power. The analysis highlighted that a wind-power future would require fundamental transformations of established distribution infrastructures and end-use domains. Heating – and especially district heating – was assigned an important role as 'batteries', capable of storing electricity in the form of hot water, that levelled out the intermittent production patterns of wind turbines.

These technical low-carbon futures created various tensions between the political decarbonisation objectives and the established governance arrangements. One problem was that the established business and socio-economic price signals were not likely to effectively promote longsighted political carbon-reduction targets. Second, the low-carbon-scenario analysis also suggested that an efficient decarbonisation of the energy system would require a coordinated choice between a bio-based and wind power-based transition pathway, a choice that was unlikely to be implemented through atomised pseudo market-investment decisions.

The discrepancy between political decarbonisation objectives and established governance arrangements became increasingly evident over time. In response to this discrepancy, planners and politicians in many urban regions began to display a more active interest in bringing heating investments in line with carbon-reduction ambitions. In 2010 – some years before the adoption of the new National Energy Act – this interest was initially highlighted by a working group comprising the Danish Energy Agency, Local Government Denmark (an interest organisation representing Danish municipalities) and representatives from individual municipalities. This working group argued that local energy planning had become increasingly 'stationary' and that the role of the municipality, as a 'reactive' project approval authority, did not ensure a 'coherent planning for larger geographical areas – and across municipal boundaries' (Danish Energy Agency and Local Government Denmark 2010, p. 4).

To promote more dynamic and coordinated planning for low-carbon energy-system transitions, the working group suggested a strategic energy-planning concept. This concept was based on the notion that municipalities should ensure more coordinated energy-system planning. The concept thus framed municipalities as active planning agents, rather than reactive project approval authorities. The working group further suggested that socio-economic prices should be recalibrated to value investments' contribution to politically formulated targets, notably decarbonisation targets. It was also argued that the unit of socio-economic analysis should not be the individual investments, but rather integrated plans for the development of entire regional energy systems. It was proposed that municipalities should be authorised to reject investments that did not comply with integrated plans for the development of the regional energy system, even if these investments could be demonstrated on an individual basis to be socio-economically optimal (Danish Energy Agency and Local Government Denmark 2010, p. 6).

While the suggestions of the working group failed to significantly influence the network of analytical assumption that informed the established market-imitating knowledge assemblage, it nevertheless inspired planners in major cities and regions to engage further in developing low-carbon heat-governance arrangements on a voluntary basis. A discussion that emerged in the context of these governance arrangements concerned the role of biomass in low-carbon futures. While the socio-economic price signals issued by the energy agency render biomass intelligible as a carbon-neutral (and hence economically having no CO_2 costs) and an unlimited resource (low future price), local planners insisted that this price signal did not reflect political concerns nor economic uncertainties. In particular, the use of imported biomass for heating was seen as problematic. A main concern was that biomass could be characterised by word-wide scarcity in the future as countries seek to decarbonise. Another concern was that available biomass should be reserved for a low-carbon transition in the transport sector, where alternative low-carbon fuels were more difficult to identify. For example, a strategic energy plan from Aalborg municipality stated that:

> All biomass resources for future use in Aalborg municipality, should as a rule not exceed the local biomass potential. From a global point of view, it will accordingly not be sustainable if Aalborg municipality, for example, has to import large quantities of biomass from neighbouring or third-world countries in order to meet the objective of a carbon neutral energy supply.
> (Forsyningsvirksomhederne 2011, p. 13 [authors' translation from Danish])

Though the local heat-governance arrangements shared the same low-carbon objectives, they varied considerably in character across municipalities. Some merely consisted in a series of general planning principles, while others were more comprehensive and detailed. One of the most comprehensive low-carbon governance arrangements was developed in the capital region. This governance arrangement, which involved a collaboration between a range of municipalities and energy utilities, aimed to decarbonise the regional energy system by 2035. A main concern was to counteract atomised investment activities. The manager thus promoted the initiative as 'planning that aims to prevent decisions that may make sense for the individual municipality or the individual utility, but that represents a clear suboptimisation from the perspective of the system as a whole' (Teknik og Miljø 2015).

To ensure coordinated investments for a low-carbon energy system, a regional low-carbon knowledge assemblage was developed. This included scenarios for coordinating investments in heating (Ea Energianalyse 2015). Similar to the established market-imitating knowledge assemblage, a main objective was to identify the socio-economic optimal investments. Socio-economic optimality was framed differently from the market-imitating knowledge assemblage. First, the object of socio-economic analysis was the regional heating system in its totality, rather than individual investment options. Second, only investment scenarios that complied with political objectives (a carbon-neutral heating system in 2035) were included in the analysis. Third, the regional investments scenarios were not analysed in the context of a status quo-based national baseline scenario, but rather they were compared against low-carbon futures developed by the Energy Agency (Ea Energianalyse 2015, p. 14). These futures operated with different assumptions regarding the amount of available biomass for heat and power production and with different assumptions regarding the future demand and price of electricity. Compared to the established market-imitating knowledge assemblage, analytical assumptions were thus recalibrated to local heat-investment scenarios within the context of an objective of low-carbon development of the broader national energy system.

The investment scenarios suggested that socio-economic optimal investments patterns in the regional heat system as a whole consist of a pervasive electrification of heat production. From 2015 to 2025, electric heat pumps that generate heat from the ambient air or from water would replace the use of natural gas, and account for approximately 20 per cent of total heat production. After 2035 the electrification would continue and was expected to account for between 60 to 70 per cent of total heat production by 2050. This energy source would be supplemented by thermal heat production from waste incineration and use of surplus heat from industry (Ea Energianalyse 2015).

Our analysis illustrates how the political revaluation of carbon emissions opened up an arena for low-carbon knowledge politics in Danish heat governance. The response of the Energy Agency, who controlled the dominant market-imitating knowledge assemblage, was to attempt to pacify and sideline concern with carbon emissions through the development of technical low-carbon energy-system scenarios that had no direct implications for the price signals used to steer the governance of heat investments. Major city regions, in contrast, exploited the concern over carbon emissions to actively criticise the market-imitating knowledge assemblage for: (1) undervaluing carbon emissions; (2) promoting uncoordinated investment decisions among technically interdependent utilities; and, (3) uncritically promoting biomass as an unlimited and carbon-neutral fuel. In the capital region these criticisms provoked the development of alternative knowledge assemblages for regional low-carbon heat transitions. This assemblage redefined heating as a low-carbon component of the future energy system, implying that only regionally coordinated low-carbon investments were acceptable investment options, and that such investment options needed to be qualified in the context of the decarbonisation of the national energy system as a whole. The high-carbon baseline future used as the standard planning future by the Energy Agency was thus rendered illegitimate by this local knowledge assemblage.

Investment and planning rationalities under multiple knowledge assemblages

Due to the co-existence of multiple knowledge assemblages, the municipalities and heating utilities involved in planning and investment decisions in regard to heating were faced with three distinct ways of knowing and valuing investments options. These were 'heat as business-economic object', 'heat as socio-economic object', and 'heat as regional low-carbon object'. So how did these different ways of knowing influence actual investment decisions and planning practices?

For the heating utilities, the business-economic valuation was typically perceived to be most overtly relevant, as the business-economic effects of investments had direct consequences for energy bills. Due to taxes that favoured biomass over other fuels, the business-economic qualification typically promoted investments in biomass boilers over, for example, electric heat pumps. Furthermore, investments in biomass boilers were sometimes actively promoted by national policymakers. In 2012, for example, a series of combined heat and power plants fuelled by natural gas ran into economic difficulties. The national parliament responded by passing a bill that allowed the 50 most expensive plants to convert from gas to biomass on business-economic grounds, regardless of their socio-economic effects (Danish District Heating Association 2014).

However, business-economic valuations were neither uncontested nor unproblematic. First, many municipal administrations had formulated decarbonisation goals and there was increasing scepticism towards increased use of biomass that were not reflected in price signals. Second, business-economic valuations were restricted by the requirement for positive socio-economic effects imposed by the heat law. For the heating utilities there were ways to work around the socio-economic qualifications. One strategy was simply to exclude socio-economically optimal options that were not commercially preferred based on technical considerations: for example, electric heat pumps that performed well in socio-economic terms might be discarded as technically infeasible. Another strategy was to bypass the heat law. For instance, the heat law did not apply to investments in co-generation plants with a power capacity of more than 25 MW; such investments were regulated by the power-provision law, which did not require socio-economic appraisal. Third, an uncertainty facing business-economic valuations was that business-economic price signals were viewed as increasingly unpredictable. Business-economic price signals were influenced by a tax structure that was said to reflect historical, rather than contemporary, concerns. For example, electricity remained heavily taxed because the generation of electricity had been very polluting in the past. This justification was challenged by the increasing amount of carbon-neutral electricity in the grid generated from wind energy. Likewise, the justification for other elements of the tax structure, such as zero tax on biomass, was increasingly questioned. An effect of these debates was that utilities increasingly lost confidence in the business-economic price signals as reflected in an interview with a manager involved in regional low-carbon heat governance:

INTERVIEWEE: You don't know where we will end in relation to the electricity tax. You don't know if biomass will remain exempt from tax. Is that what we are going to expand? Is it something else? Is it the interconnections?

INTERVIEWER: Ok, so actors are simply unsure of the economic conditions for business decisions?

INTERVIEWEE: Yes, the uncertainty means that you are holding your investments back. And saying, is anything happening. And you may say that the present administration excels in taking things up, and not coordinating them with other things. And suddenly something's come up and they say 'now we are going to do it in this way'.

While the business-economic prices signals provided a valuation that was overtly relevant to the utilities, this valuation was seen as increasingly uncertain and was furthermore challenged by political objectives on biomass-independent decarbonisation.

Compared to business-economic valuations that were typically prioritised by the utilities' focus on energy bills, the socio-economic valuation was dominant within the planning and policy community. Socio-economic price signals for example constituted the default economic input to energy models used in scenarios concerning the future development of the energy system, as these price signals were seen as efficient from a societal point of view. In contrast to the business-economic valuation, the socio-economic price signals typically prioritised investments in the electrification of heating over investments in biomass boilers, as these prices omitted the tax inducements favouring biomass. In actual investment practices, the socio-economic price signal was not very successful in promoting electrification of the system because this was unfavourable from a business-economic perspective. In contrast, socio-economic valuation played a pivotal role in low-carbon investments in heat-distribution infrastructures. Such low-carbon investments mainly consisted of replacing established gas pipes with low-carbon district-heating pipes. To protect their market, gas utilities often challenged such conversion projects on the ground that they did not generate positive socio-economic effects. Oftentimes the gas utilities scrutinised socio-economic calculations in order to identify individual neighbourhoods or individual industrial facilities where a conversion did not generate socio-economic benefits in order to have them taken out of the project. Controversies in relation to socio-economic calculations could arise, for example, over the expected energy efficiency and maintenance costs of private gas boilers or over the expected number of connections to newly established district heating infrastructure. Socio-economic costs ascribed to natural gas and carbon emissions were experienced by the district heating utilities as a barrier to long-term low-carbon investments in district heating infrastructures. A manager from a waste incineration and district heating company in the capital region described this issue in the following terms:

INTERVIEWEE: When we are applying for approval of a district heating project we are looking into the framework specified by the heat provision law. When we then look at where society should be heading (…) then we turn to the political visions about carbon neutrality for heating and power in 2035 (…)

INTERVIEWER: How well do these two futures go together?

INTERVIEWEE: Very poorly!

In response to the calculative difficulties in making conversion projects appear socio-economically beneficial, the Danish District Heating Association became increasingly critical about the use of the European CO_2 quota price as socio-economic cost and advocated that the CO_2 prices needed to be drastically adjusted from US$9/ton CO_2 to US$175/ton CO_2 (Danish District Heating Association 2017).

Finally, in urban areas like the capital region, investment options could also be valued as components of a regional low-carbon heating system. The challenges of translating regional low-carbon investment scenarios into actual investment decisions were considerable, as investment options proposed in these scenarios typically conflicted with business-economic valuations, and because objectives to limit the use of biomass conflicted with the socio-economic price signals issued by the Energy Agency. In the capital region ideas to translate their low-carbon heat-investment scenario into operational investment plans for individual heating utilities had to be abandoned. The low-carbon investment scenario served a more indirect purpose. It was used by municipal energy planners to justify politically appointed board members actively promoting low-carbon investment strategies. This, in turn, provoked the management within the heating utilities to complement their traditional focus on daily operation and optimisation of heat provision, with a strategic focus on long-term low-carbon investments. An effect of such strategic capacity building was a more frequent interaction among utilities and municipalities. In the Copenhagen region, this type of interaction resulted in attempts to develop new types of investment partnerships. For instance, a partnership was formed to investigate the possibility for a joint investment among several utilities in a large-scale heat-storage facility in order to increase overall regional heat-system flexibility (VEKS and Høje Taastrup Fjernvarme 2017).

The co-existence of multiple knowledge assemblage for qualifying investment thus situated actors involved in actual planning and investment activities in complex and sometimes contradictory situations. A common way to organise this complexity was to divide it into different temporal orientations. Strategic long-term investment perspectives were thus increasingly qualified with reference to low-carbon knowledge assemblages, whereas shorter-term investments were justified mainly in terms of business-economic price signals.

Discussion and conclusion

The aim of this chapter has been to highlight how actors involved in everyday heat governance in Denmark are dependent on elaborate knowledge assemblages for making their reality comprehensible and actionable. More specifically we have illustrated how heat provision is rendered governable through the construction of calculative spaces made up of different elements such as price signals and benchmark scores.

We have furthermore argued that such knowledge assemblages should not be mistaken for politically neutral decision-support tools. This is because such assemblages are only able to make complex and uncertain realities and futures knowable in a relatively consistent and unequivocal way by means of a complex network of contentious analytical assumptions. Knowing is thus a fundamentally political ingredient in contemporary system governance because the networks of analytical assumption that are needed in constructing stabilised fact-like knowledge have political repercussions for how actors are able to conceive challenges, value alternatives and make sense of their own agency. Based on this knowledge-politics perspective, we can identify a number of key questions pertaining to contemporary heat governance: How are particular knowledge assemblages constructed? What do they make visible and valuable? Who polices them? And, how can such assemblages be transformed?

To address these questions, we have discussed a contemporary knowledge controversy among a dominant market-imitating knowledge assemblage, controlled by the Danish Energy Agency, and a more recent – and, so far, less influential – regional low-carbon knowledge assemblage. Our analysis illustrates that the dominant market-imitating knowledge assemblage so far leaves little room for alternative knowledge assemblages to influence political rationalities and practical governance activities. This, we suggest, is the case because the analytical assumptions that inform the dominant market-imitating knowledge assemblage enjoys the support of a powerful configuration of expert communities and political interests. The market-imitating knowledge assemblage is thus closely calibrated with a hegemonic economic knowledge regime that has been developed and promoted by the Ministry of Finance over several decades. The central focus of this broader knowledge regime is to optimise the competitiveness of the Danish economy by decreasing public expenditures and by increasing the supply of labour. Under this knowledge regime, low-carbon policies and investments are valued as an impediment to the competitiveness of the national economy. The low-carbon knowledge assemblages are thus competing for political salience in the context of a hegemonic regime in which low-carbon policies are viewed as an impediment to the competitiveness.

Low-carbon political rationalities are thus likely to presuppose a long-term struggle to reorganise an interwoven complex of political interests and expert communities that currently controls the analytical assumptions through which knowing is being controlled and made possible.

Notes

1 Department of Planning, Aalborg University, Denmark.
2 Department of Planning, Aalborg University, Denmark.

References

Bueger, C., 2015. Making things known: epistemic practices, the United Nations, and the translation of piracy. *International Political Sociology*, 9 (1), 1–18.

CarbonBrief, 2017. Explainer: the challenge of defining fossil fuel subsidies. COWI, 2009. *Analysis of the incentive structure in the district heating sector – final report* [*Undersøgelse af incitamentsstrukturen i fjernvarmesektoren Endelig rapport*].

Danish District Heating Association, 2014. The most expensive empty field plants have cut 6700 dkk of the heat price [De dyreste barmarksværker har skåret 6700 kr. af varmeprisen].

Danish District Heating Association, 2015a. *The heat provision law and the project consolidation act in relation to heat projects and local heat planning* [*Varmeforsyningsloven og projektbekendtgørelsen i forbindelse med varmeprojekter og varmeplanlægning lokalt*].

Danish District Heating Association, 2015b. *Economy models in district heating projects* [*Økonomimodeller i fjernvarmeprojekter*].

Danish District Heating Association, 2017. *Socio-economic CO2 price based on damage effects rather than supply and demand* [*Samfundsøkonomisk CO 2-pris baseret på skadevirkning frem for udbud og efterspørgsel*].

Danish Energy Agency, 2014. *Energy scenarios towards 2020, 2035, and 2050* [*Energiscenarier frem mod 2020, 2035 og 2050*].

Danish Energy Agency, 2016. *Regulation and planning of district heating in Denmark.*

Danish Energy Agency, 2017. *Socio-economic assumptions for energy prices and emissions* [*Samfundsøkonomiske Beregningsforudsætninger for Energipriser Og Emissioner*].

Danish Energy Agency, 2018. *Baseline projection 2018* [*Basisfremskrivning 2018*].

Danish Energy Agency and Danish Competition Authority, 2004. *Framework conditions for the district heating sector* [*Rammevilkår for fjernvarmesektoren*].

Danish Energy Agency and Local Government Denmark, 2010. *Presentation regarding strategic energy planning* [*Oplæg om strategisk energiplanlægning*].

Ea Energianalyse, 2015. *Analysis of the Capital Regions future heat supply* [*Analyse af hovedstadsregionens fremtidige fjernvarmeforsyning*].

EcoGrid eu, 2016. *EcoGrid EU – A Prototype for European Smart Grids. Deliverable D6.7 Overall evaluation and conclusion.*

Euractive, 2018. IEA scenarios 'inconsistent' with Paris climate goals, study warns.

Forsyningsvirksomhederne, 2011. *Energy Strategy for Aalborg Municipality till 2030* [*Energistrategi for Aalborg Kommune frem til 2030*].

International Energy Agency, 2016. *World Energy Outlook 2016*. International Energy Agency.

International Energy Agency, 2017. *World Energy Outlook 2017*. London: OECD.

JRC Science Hub, 2017. *Smart grid projects outlook 2017 Facts, figures and trends in Europe*.

Lund, H., Werner, S., Wiltshire, R., Svendsen, S., Thorsen, J.E., Hvelplund, F. and Mathiesen, B.V., 2014. 4th Generation District Heating (4GDH). *Energy*, 68, 1–11.

Miller, P. and Rose, N., 1990. Governing economic life. *Economy and Society*, 19 (1), 1–33.

Mitchell, T., 2008. Rethinking economy. *Geoforum*, 39 (3), 1116–1121.

Stiglitz, J., Stern, N., Duan, M., Edenhofer, O., Giraud, G., Heal, G., Lèbre la Rovere, E., Morris, A., Moyer, E., Pangestu, M., Shukla, P., Sokona, Y. and Winkler, H., 2017. *Report of the High-Level Commission on carbon prices*. Geneva: ipcc.

Teknik og Miljø, 2015. Mayors wants binding collaboration [Borgmestre vil have forpligtende Samarbejde], 20 Jun, p. 2.

The Climate Institute, 2017. *Factsheet – Why International Energy Agency scenarios should not be misinterpreted as forecasts*.

The inter-ministerial working group concerning work regarding comparison of efficiency in the district heating sector, 2015. *Modernised regulation in the district heating sector* [*Moderniseret Regulering I FjernVarmesektoren*].

VEKS and Høje Taastrup Fjernvarme, 2017. *Heat storage in Høje Taastrup* [*Varmelager i Høje Taastrup*].

Voß, J.-P. and Freeman, R., 2016. *Knowing governance: The epistemic construction of political order* 1st ed. London: Palgrave Macmillan.

7 Urban metabolism as governmentality

Governing the city of flows

Vanesa Castán Broto and Louise Guibrunet

Introduction

What makes a city? As nodes in the 'physical economy', cities mediate the transformation of resources that support contemporary production and consumption systems (Schandl *et al.* 2016). Kaika (2005, p. 5) described a 'city of flows' built upon a scaffolding of 'hidden material networks and flows that urbanize nature'. Among concerns about ensuring sustainability and securing resources, managing the material fabric has become a central problem for contemporary cities (Swilling *et al.* 2011). Urban metabolism is a rapidly developing epistemic tool to govern urban flows. This chapter analyses how the deployment of urban metabolism supports the formulation of new political rationalities for urban governance. Foucault described the art of government as the control of people and conduct (Dean 2009). He coined the notion of governmentality to explain the rise of an expansive form of government that depended on specific forms of rationality, or techniques of government, for the definition and control of populations. As Kaika and other political ecologists have shown (Heynen *et al.* 2006; Gabriel 2014), these forms of expansive authority are built through attempts to control land and resources. They require not only providing adequate means to securitise the resource base for the reproduction of the urban economy but also ensuring the means of control of resources to support capital appropriation by the urban elites (Swyngedouw 1995).

Sustainability transitions result from broad processes of socio-technical change that lead to the reconfiguration of urban infrastructure regimes (Monstadt 2009). Sometimes the transition is predicated on attempts to challenge the 'conduct of conducts', that is, changing the mechanisms of population control. While *et al.* (2010), for example, have explained how concerns with climate change have generated different mechanisms for carbon control, in particular, a carbon calculus whereby actions can be measured against the reduction of carbon emissions. Alongside this carbon calculus, there are also attempts to territorialise carbon emissions. For example, at the city level, inventories of carbon emission help to allocate emissions to different activities that can then be regulated through different regulatory mechanisms or changes in service provision (Rice 2010).

This process is not exclusively confined to carbon control and attempts to reduce emissions. The aspiration to achieve sustainable cities pertains to different flows of natural resources at multiple levels. In each case, there is a need to find the means to understand how resources operate in cities. As the editors of this volume explain, 'urban systems are not self-evident, with naturally defined boundaries, functions and sustainability challenges' (Chapter 1, p. 2). Thus, a crucial part in governing transitions is finding the means to calculate and make visible the processes of transformation of resources to make it possible to define potential actions, means of intervention and possibilities of success.

Following this Foucauldian line of thinking, urban metabolism is to urban resources what statistics are to populations: a means to build authority through the definition of the problem and the designation of a programme of intervention. Urban metabolism has revolutionised thinking about resource use in cities. While the notion of metabolism can be employed differently to analyse the politics of the capitalist economy (Heynen *et al.* 2006; Castán Broto *et al.*, 2012), the dominant understanding of urban metabolism involves creating a systematic way of understanding and documenting resource use. A central tenet of urban metabolism is to conceptualise the city as constituted by resource flows (although the notion of stocks is also gaining currency). Within the field of industrial ecology, for example, urban metabolism entails the practical application of Material Flows Analysis (MFA), which uses existing statistical information on the use of resources to quantify resource flows associated with the city's consumption of natural resources and generation of wastes (see, for instance: Wolman 1965; Kennedy *et al.* 2007; Barles 2009; Bristow and Kennedy 2013). This conceptualisation is used to design actions to optimise the urban resource cycle and improve resource management (Kennedy *et al.* 2011; Alfonso Piña and Pardo Martínez 2013; Rosado *et al.* 2014). The central idea is to achieve a change from linear metabolisms to circular ones (Girardet 2004). In this way, the diagnostic method (MFA) is linked to a clear programme of action that enables gaining authority over the city.

This chapter presents an exploration of the concept of urban metabolism and MFA as tools to govern the sustainability transition in cities. Research on urban metabolism has tended to examine the fit of MFA methods with the actual city (see, for instance: Barles 2007; Moore *et al.* 2013; Hoekman 2015). More recently, as a result of debates on urban political ecology, there has been a realisation that urban metabolisms depend on particular configurations of power. The result has been a call for the incorporation of such political concerns in urban-metabolism thinking (a call for a politico-industrial ecology) (Newell and Cousins 2014; Newell *et al.* 2017). What we offer here is a reinterpretation of urban metabolism as a tool for governing the city through the configuration of knowledges, visibilities and rationalities. In doing so, we are inspired by a Foucauldian analysis of MFA as constituting a new governmentality that is changing ideas about contemporary urban management.

We focus on the case of the metabolism of solid waste in Mexico City. Mexico City is an exemplary case in the use of MFA to manage an urban sector.

The Ministry of the Environment in Mexico City has been using MFA methods to produce annual 'waste inventories' during the past decade, which have been used to produce the two most recent waste management programmes, in 2010 and 2016 (Gobierno del Distrito Federal 2010; SEDEMA 2016). In this context we inquire how the deployment of urban metabolism renders the urban fabric in Mexico City governable, the kind of urban relations that it makes visible and the ways in which urban metabolism reproduces urban inequality.

Following these questions, we examine everyday realities of waste management, focusing on dilemmas that emerge within current policy-making. Alongside the quantified and formal spaces of waste management, we observe some parallel informal waste-handling activities not fully recognised in waste policy. We also represent the city's waste metabolism spatially, to uncover spatial inequalities in waste management practices. In particular, the case raises concerns about the possibility of addressing resource sustainability while also delivering environmental justice (Allen and You 2002; Swilling and Annecke 2012; McFarlane 2013). There is an increasing body of research that has looked into extending the urban metabolism framework to incorporate inequalities in the distribution of resources (Royden Turner 2012; McFarlane 2013; Smit *et al.* 2017). In practice, this body of evidence has only had a limited influence on the practice and conceptualisatin of urban metabolism.

An overview of the intellectual and methodological concerns of urban metabolism

Urban metabolism draws on Materials Flows Analysis (MFA) or similar methodologies (Zhang 2013). The concern is to develop programmes of intervention that improve resource efficiency at different scales (neighbourhood – industrial or residential – city, country, region) (Fischer-Kowalski 1998; Baccini and Brunner 2012; Hammer *et al.* 2003). For instance, using wastes as inputs in industrial or productive processes is thought to improve circularity (Girardet 2004). Alternative solutions propose substituting inputs through local production (for instance, local production of energy or rainwater harvesting). The field of industrial symbiosis and the zero-waste movement exemplify the kind of interventions proposed in MFA methodologies.

MFAs use statistical datasets as the primary source of data. Because of this, the scales of analysis reflect administrative units, such as cities' administrative boundaries (Baccini and Brunner 2012; Ferrão and Fernández 2013). As the focus is on flow quantification, the analysis excludes typically any flow that cannot be quantified or for which the data is not considered robust (Currie *et al.* 2015). Likewise, MFA analyses with a spatial component are rare (exceptions being Chrysoulakis *et al.* 2013; Cousins and Newell 2015).

As analyses conducted to date have primarily drawn from officially published statistics, this means that urban metabolism reflects the quantity and quality of official data on resource management. Processes that are not monitored by the state (such as those within the informal economy) hardly feature in urban metabolism analyses (Currie *et al.* 2015; Hoekman 2015). Moreover, urban metabolism studies tend to be macro-studies that provide trends as to cities' overall energy and material consumption, particularly in the case of urban Latin America (see, for instance: Delgado Ramos *et al.* 2012; Alfonso Piña and Pardo Martínez 2013; Kennedy *et al.* 2015). They rarely engage with the political implications of such trends, particularly regarding unequal access to resources by different households, or the differentiated environmental impacts for urban dwellers.

New waste industries coexist (and compete) with citizen-led practices of waste management (Scheinberg *et al.* 2011). Waste management is a source of livelihoods for some households: many waste handlers earn a living by collecting, selecting, transporting and recycling waste (Wilson *et al.* 2006; Linzner and Ali 2013). The literature on informal waste management in a Latin American context is vast. Scholars have documented the role waste management plays in supporting livelihoods (Florisbela dos Santos and Wehenpohl 2001; Hartmann 2012); its governance system (Medina 2000); the extent to which illegal practices permeate both informal and formal waste management (Castillo Berthier 2003); and its health implications for waste handlers (Wilson *et al.* 2006).

Urban metabolism principles are relevant to the waste sector (Desrochers 2002; Chertow 2008). Sustainable waste management combines a concern with reducing natural resource consumption, producing less waste, and opening up economic opportunities. For example, the management of solid waste through recycling goes hand in hand with the development of the recycling industry. We focus here on a less understood aspect of urban metabolism, in terms of the rationalities that are mobilised in the waste management sector and their consequences.

Waste management in Mexico City

Mexico City is the capital of Mexico and one of the most populated cities in Latin America. Mexico City has high levels of waste generation: 1.5 kg per inhabitant per day, which is above the world average of 1.2 (Hoornweg and Bhada-Tata 2012). As explained above, Mexico City Ministry of the Environment is a pioneer in the practical use of an MFA to develop urban waste management policies.

Mexico City is, administratively, one of the 32 states of the United States of Mexico, which is a federation. Approximately 8.8 million people lived in that state as of 2010 (INEGI, 2010). The city has expanded beyond its administrative boundaries: the metropolitan area is composed not only of Mexico City but also of over 50 municipalities in the neighbouring states: the State of Mexico and the State of Hidalgo (see Figure 7.2). In total, nearly 22 million people inhabit the metropolitan area (SEDESOL *et al.* 2012). This metropolitan area is defined as a commuter shed (defined by daily commuting patterns) but lacks any governing entity.

Waste management responsibilities are shared among the three levels of the Mexican government. At the federal level, the Ministry of the Environment and Natural Resources sets the legal responsibilities and environmental standards that are enforced at the local level, through the *General Law for Waste Prevention and Integral Management*. Also, the *Law of Climate Change Mitigation, Adaptation and Sustainable Development* defines emissions thresholds for waste management and disposal processes. State governments design and implement waste management strategies: Mexico City's *Law of Solid Waste* thus sets the rules for urban waste management, which includes the regular publication of a *Programme for the Integral Management of Solid Waste*. This Programme – last published in 2016 (SEDEMA 2016) – focuses on measures that can improve the operation of the waste management system. The programme's strategies are informed by a waste inventory that applies techniques of MFA. The inventory contains annually updated data and visualisations of waste flows in the city, both for policy-making and for communication to the wider public.[1]

Through its waste management programme, Mexico City's government determines the logistics and strategies for waste management. The State's Ministry of the Environment (SEDEMA in Spanish) delivers the city-wide waste management strategy and also sets standards for waste management operations. In this way, it regulates the operation of private waste-collection companies. The State's Ministry of Urban Services and Public Works (SOBSE in Spanish) operates the waste management infrastructure, including the Municipal Waste Transfer Centre (where waste is separated before reaching disposal sites), composting sites, waste selection plants (which is where recyclable material is extracted) and landfills. The municipalities are in charge of street cleaning and domestic waste collection.

Research methods

This chapter presents a critical analysis of Mexico City's waste inventory, reflecting particularly on the alignment of diagnosis with governance proposals, and the acknowledgement of waste management dilemmas excluded from current assessment and calculation. To do so, we are offering a reinterpretation of a case study presented originally in a comparative paper on urban metabolism (Guibrunet *et al.* 2016). The study is based on a six-month fieldwork period in Mexico City, where 58 interviews (semi-structured and unstructured) were conducted with civil servants and formal and informal waste workers. Also, our assessment of urban metabolism includes an analysis of the 2015 inventory (SEDEMA 2015), which is the one used to inform the 2016 waste management programme.[2] A qualitative analysis of in-depth interviews with civil servants involved in waste management ($n = 6$) complements the document analysis.

The results are presented as follows: first, we provide an overview of the information presented in the Inventory and how they relate to a particular programmatic approach to waste management. Then, we examine three dilemmas overlooked in the analysis. In each case, we contrast the results from the inventory with our observations on the ground. The first dilemma relates to the conflicts resulting from the unequal spatial distribution of waste-disposal facilities at the metropolitan level.

The second dilemma relates to the relationship between socio-economic inequality and the quality of waste-collection services at the neighbourhood level. To examine this issue, we focus on a single municipality within Mexico City, called Cuauhtémoc. To measure socio-economic inequality, we use the marginalisation index produced by the National Council for Population, which divides the national population into five categories (very low, low, medium, high and very high marginalisation) (CONAPO 2010). As this index is calculated at the national level, finding populations with a very low marginalisation and others with a high marginalisation within the same neighbourhood is an indication of the high socio-economic inequality concentrated in a small territory. To estimate the quality of waste-collection service we develop a calculation based upon primary data obtained during interviews with Cuauhtémoc civil servants. The dataset is made up of the number and volume of illegal dumps found in each neighbourhood, which reflects a lack of municipal collection. Civil servants use this data as a proxy to estimate the quality of the service. We have adapted this data to provide an index of the quality of the waste-collection service, by multiplying the number of illegal dumps by the volume of illegal dumps. This is then normalised by dividing it by the area of the neighbourhood. We then classified the data in three levels of quality (good, medium, bad), based on the Jenks Natural Breaks algorithm (the algorithm suggests that natural breaks between classes are based on natural groupings inherent in the data, where similar values are grouped and the differences between classes are maximised).

The third dilemma relates to the different waste management systems that operate in Mexico City in parallel to the formal government operation. The objective here is to shed light on the flows of waste that are overlooked in the Inventory. This is based on ethnographic interviews with institutional and informal waste handlers throughout the Cuauhtémoc municipality ($n = 45$).

Territorialising urban metabolism

Objectives of the waste inventory

The main component of the Inventory is a quantitative waste-flows diagram, representing domestic and commercial waste flows (industrial and construction wastes are excluded) from their source to their final disposal site. It provides a visual representation of waste flows in Mexico City as they are reflected in official statistics. It also represents the actors who intervene in handling waste, including those driving the waste-collection trucks, and those working at the waste-transfer centres and the waste-selection plants. The main objective of the Inventory is to identify the source and end point of flows and quantify them. On the other hand, the document lacks an analysis of the adequacy of the knowledge base, the quality of the data, or alternative framings of the problem. There is no written assessment of the uncertainties inherent to its elaboration.

Figure 7.1 provides an overview of the data in a Sankey diagram, a standard method for the visualisation of urban metabolism. The system manages 12,893 tons of waste per day. The majority of the waste is produced by households (48 per cent). Other significant sources of waste are shops (15 per cent), services (15 per cent) and markets (14 per cent). A small fraction of hazardous waste (2.7 per cent) is managed separately. Additional waste from the neighbouring State of Mexico is also managed in the waste-selection plants (1,071 tons per day).

Around 58 per cent of the waste is disposed of in landfills. Cañada and Milagro are the largest disposal sites for Mexico City, both receiving over 3,000 tons of waste per day. Another site, in Cuautitlán, receives over half that amount. Two smaller sites are located in Cuautla and Tepozotlán. Some of the waste is composted (12 per cent) or recycled (3 per cent). Another 12 per cent of the waste is selected by the municipal workers to be sold as recyclable materials (this is the flow labelled 'waste-picking in government infrastructure'). The remaining 15 per cent is unaccounted for (although an interview with a civil servant reveals that this number represents the waste-picking taking place on the municipal truck – before reaching the waste management infrastructure).

This data is partial, as it only takes into account waste that is managed by government workers and infrastructure. This includes waste collected by municipal employees or by the city's urban services department; and waste that is disposed of in government-managed composting sites, recycling schemes or landfills. Any waste flows not managed by governmental actors (either by formal private waste-collection companies or informal waste-handlers) are absent in the Inventory.

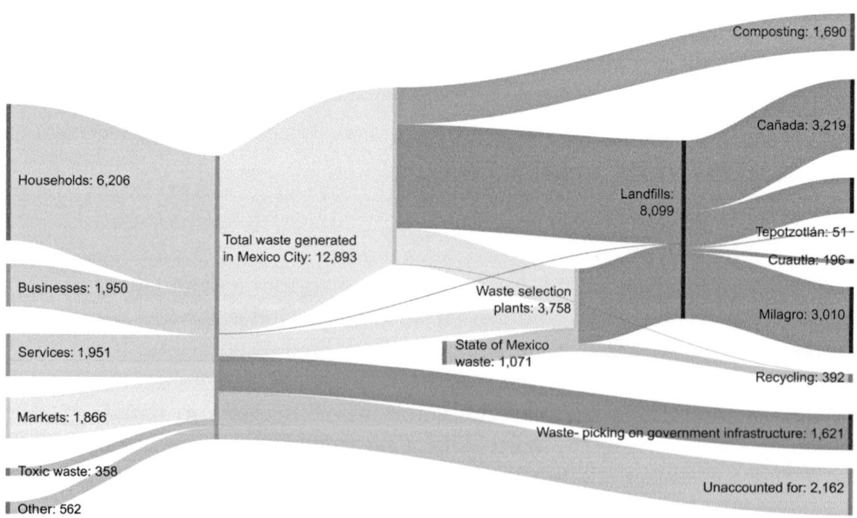

Figure 7.1 Sankey diagram of waste flows in Mexico City, in tons per day.

Source: authors' own elaboration with data from SEDEMA 2015.

Civil servants know of the existence and importance of those unofficial waste flows: they mention them in the interviews. Why are these flows absent? We find an explanation in the process of production of the Inventory. In the interviews, government officials explain this exclusion by the limited resources available for data collection and analysis. Indeed, the institution that produces the Inventory (SOBSE) has no responsibility or budget for data collection beyond its own operation. This reflects an inadequate problem-framing.

Knowledge-production processes always depend on the delimitation of the problem of study. In this case, however, the definition of objectives marks priorities for data collection. The objective of the Inventory is to support the development of a governmental strategy, the main objective of which is, according to a government official, 'to dispose of waste in the cheapest possible way'. Any flow that is not relevant to that objective, an interviewee explains, is of no interest in the Inventory. The Inventory focuses on governmental flows because those are the flows that can be controlled, and that therefore impact on the government's budget (in terms of running the government infrastructure).

The Plan is designed by SOBSE, and proposes strategies of action for SOBSE alone. SOBSE is in charge of finding and paying for landfill sites. Reducing waste to landfill is thus the Plan's primary operative goal. To this end, the Plan proposes processing organic waste in composting plants and bio-digesters, and non-organic waste in incinerators, to produce energy out of the heat (SEDEMA 2016). These proposals will also create investment opportunities, as public–private partnerships will be the main form of organisation for the new infrastructure. This mode of planning through directing investment is in line with SOBSE's capacities as the Ministry of Public Works.

From the interviews, it appears that SOBSE representatives have little or no interest in alternative waste management strategies. These could include waste-prevention measures (such as prohibiting the sale of water bottles or plastic bags in the city) and recycling measures (enhancing separation procedures). Informants explained that implementing these measures is not within SOBSE's competences, and thus cannot form part of the new Plan.

Thus, the framing of the Inventory reflects the lack of integrated planning in urban environmental governance. The separation of government institutions leads to a sectoral treatment of environmental problems, such as waste management. Such sectoral treatment means that some key aspects of waste management – those which lie beyond the competencies of the institution in charge of the Plan – are overlooked. Planning proposals are directed towards specific areas without an effective treatment of the waste system as a whole.

Simultaneously, the Inventory reflects how urban metabolism assessments, such as MFA, are shaped by the contextual politics in which they are delivered. The following section provides an overview of the blind side of urban metabolism – the aspects of urban life that it makes invisible – as it is manifest in Mexico City. We have identified three crucial aspects of waste management that are not covered in the Inventory, as a result of its focus on a traditional MFA; these can be explored by using alternative (qualitative) methods of researching material flows. The first blind side is the lack of awareness of the spatial organisation of waste flows. The second relates to the scale of representation of those flows. The third one is the exclusion of flows for which data is not available. Each of these aspects is developed in turn.

Defining the spatial organisation of waste flows

The Inventory responds to a narrow conceptualisation of the spatial distribution of waste flows in Mexico City. It fails to show that the environmental impacts of the waste management system concentrate outside the city borders, in the neighbouring states of Mexico and Morelos. The spatial structure of the city shapes patterns of waste circulation. After collection, the domestic waste is first sent to the Municipal Transfer Stations in municipal waste-collection trucks. In the Stations, workers separate the organic waste fraction suitable for composting and the waste with recyclable potential. Both waste flows continue to facilities for waste recovery (composting sites and recycling schemes). The waste without composting or recycling potential is sent to the landfills instead.

Figure 7.2 maps the sites of intermediary and final disposal of waste. A spatial pattern is visible. Composting sites and waste-selection plants are infrastructures that extract value from waste. They are mostly located within the administrative boundaries of Mexico City (only two compost plants are outside, one being located at the airport, which is federal land). Landfill sites, in contrast, have negative economic and environmental impacts for the area in which they are located. They are all located beyond the administrative boundaries of the city.

This pattern is neither coincidental nor without consequences. The export of Mexico City's solid waste to its metropolitan area (mainly in the State of Mexico) causes political tensions. In the spring of 2016, for example, the 'waste crisis' made it onto the front page of national newspapers. The waste crisis emerged due to political tensions between Mexico City and its neighbouring states. During the crisis, the State of Mexico refused to receive waste in its landfills. This overwhelmed the Municipal Transfer Stations. As a result, Mexico City's government had to pay 2 million pesos (US$107,000 or €95,240 as of 2016) a day for alternative waste disposal in the Cuautla landfill, in Morelos, until the crisis was resolved; and in order to do so, it had to back down on environmental policies that the State of Mexico were contesting (Salinas Cesáreo 2016). The event underscores the dependence of Mexico City on its neighbouring states and their waste management facilities. Arguments about the spatial distribution of environmental impacts surface regularly and are not limited to waste – indeed, the waste crisis was embedded within a wider air-pollution crisis. Urban metabolism relies on a particular conceptualisation of scale that needs to be adjusted and readjusted depending on the social and political manifestations of environmental conflicts.

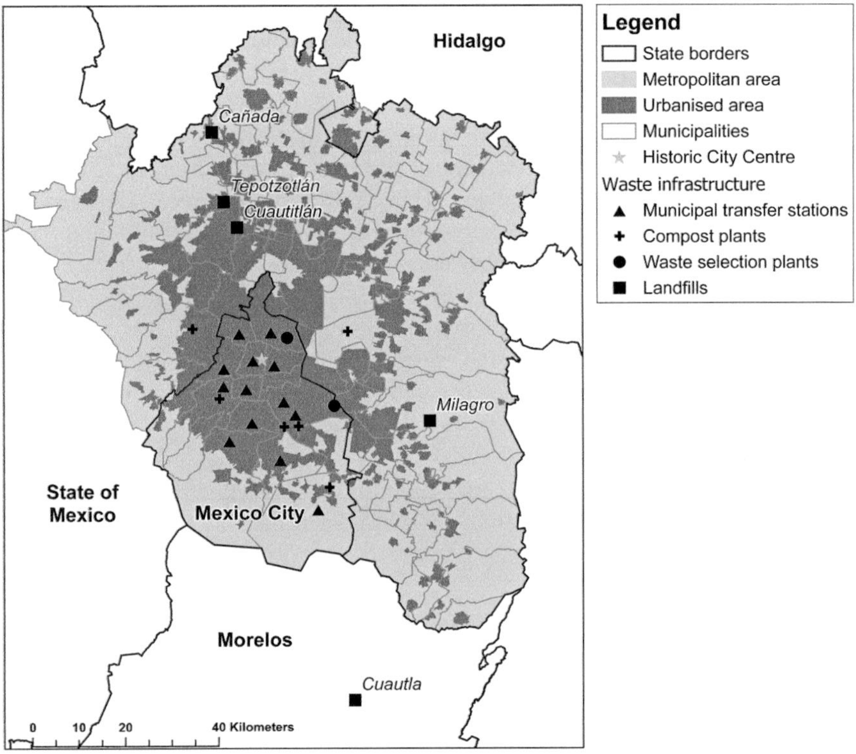

Figure 7.2 Location of Mexico City's waste management infrastructure in the metropolitan area of Mexico City.

Source: author's own elaboration with data from SEDEMA 2015; SEDESOL *et al.* 2012.

The multi-scalar nature of waste management

Following the competences of the Environment Ministry, the Inventory analysis relates to data produced at the city-wide scale. Multi-scalar relations, however, determine waste governance in Mexico City. Environmental impacts manifest with varying consequences at different scales: the previous section demonstrated the tensions between the city and its hinterland. Also, the Inventory does not zoom in to neighbourhoods; it has therefore limited capacity to explain intra-city differences in waste management patterns. This has crucial consequences in terms of the conclusions that can be drawn from the analysis.

The definition of the MFA not only excludes external instances of flows and resource sinks but also, it simplifies the internal structure of those flows. For example, the Inventory overlooks the different quality of the waste management service between neighbourhoods. There are differences between municipalities but also within municipalities in the city. This is so even though waste collection is organised at the municipal level. In contrast, the Inventory relates a homogenised vision of environmental management that limits targeted interventions.

A specific case illustrates the evident differences in service quality within a single municipality. The Cuauhtémoc municipality is in the centre of Mexico City: it contains the historic city centre, and hosts most of Mexico City's governmental offices. It is crossed by two of the main urban corridors, Reforma and Insurgentes. Cuauhtémoc has a population of over 53,000 inhabitants. Their socio-economic background is highly unequal.

Figure 7.3 compares an index of socio-economic level (the marginalisation index) with the quality of the waste-collection service. The marginalisation index shows that this is an area with extreme levels of socio-economic inequality. The northeast area of the municipality exhibits medium to high levels of marginalisation. The western side, in contrast, scores very low on the marginalisation index.

Figure 7.3 shows that there are different levels of quality of the waste-collection service in different neighbourhoods within the municipality. The

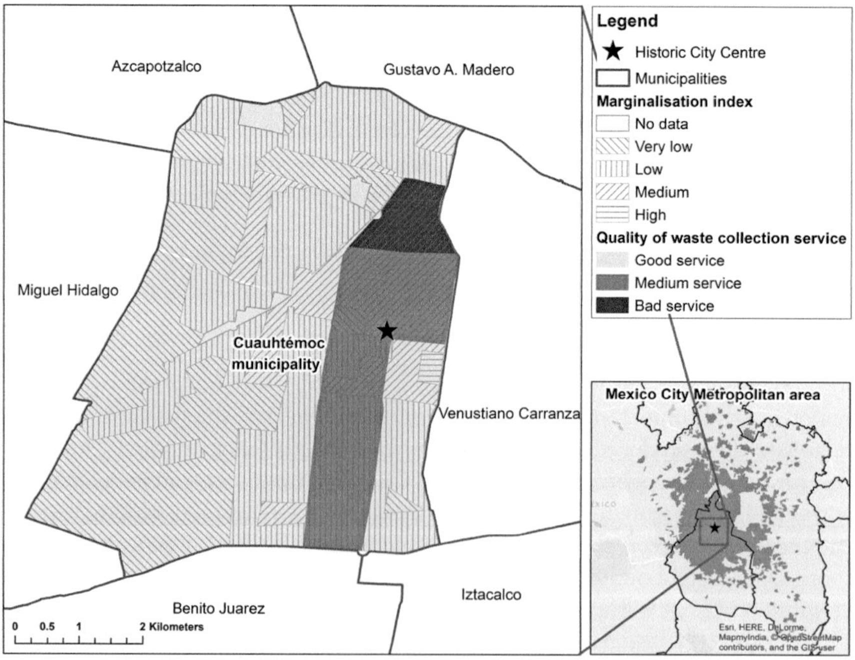

Figure 7.3 Map of the Cuauhtémoc municipality in Mexico City, showing the quality of the waste-collection service.

Source: authors' own elaboration with additional data from CONAPO 2010.

difference between an area with a good service and one with a bad service is enormous. A neighbourhood with a good service has less than five illegal dumps of waste in the streets. A neighbourhood with a bad service has over 50. The range of volumes observed goes from 600 to 48,000 kilos of waste.

As these illegal dumps are incrementally formed by the bags of domestic waste citizens drop on the street corners when they have no alternative (no adequate waste-collection service), the presence and size of illegal dumps reflect the capacities of the different municipal teams to collect all the domestic waste produced in their neighbourhood. This difference in capacities at the neighbourhood level is arguably a consequence of the uneven distribution of municipal resources (staff, trucks, financial and technical resources) between those teams. In our interviews, municipal workers and their on-the-ground managers readily recognise inequalities in service provision. In contrast, a representative of the municipality's Office of Urban Services claims that the distribution of resources is equitable. The Inventory – the official assessment of waste management in the City – offers no evidence to contradict the municipality's claim. However, a transect from the west to the eastern areas of the municipality demonstrates the physical differences that the Inventory does not reflect. The overlay of the socio-economic index with the quality-of-service index shows that waste services tend to be worse in areas corresponding to medium and high levels of marginalisation. The waste management service thus reproduces existing inequalities within the Cuauhtémoc municipality.

Going beyond official statistics and quantifiable variables

The Inventory takes into account the flows that go through the government's infrastructure (the municipal transfer stations and waste-selection plants). There is ample evidence that this assessment is incomplete as some waste management actors are absent from the waste inventory.

Figure 7.4 represents a scheme of the multiple waste flows that operate in the city. While the only flows present in the waste inventory are those labelled 'municipal collection system', Figure 7.4 shows that these represent a partial view of waste flows through Mexico City.

The first omission is the private sector. In Mexico City, municipalities cannot legally grant concessions to private companies to run the waste-collection service. The private sector still plays a role. First, private companies are hired to manage hazardous waste. Although in the Inventory hazardous waste represents about 2.7 per cent of the domestic waste (SEDEMA 2015), it could be a higher proportion when taking into account industrial waste excluded from the Inventory. Also, private companies can provide waste-collection and management services to private entities for a fee. Usually, they provide a higher-quality service than the municipal collection service. Their clients tend to be high waste generators, such as shopping centres or gated communities. Accommodated classes, which already enjoy a better quality of municipal service, may have access to this additional service. There is no data

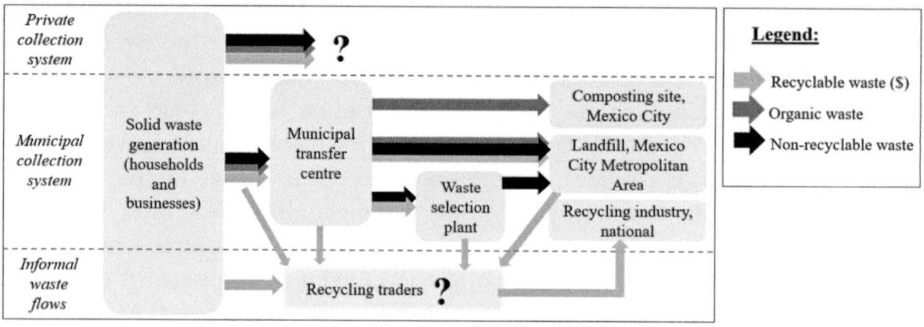

Figure 7.4 Flows of domestic waste in Mexico City – the state of knowledge.

Source: authors' own elaboration.

publicly available on the amount of waste collected in this way or the number of companies providing this service in Mexico City. This is another example of the reinforcement of inequalities in service provision, and their invisibility in the Inventory.

The other important group of non-governmental actors absent from the Inventory is informal waste handlers. The phrase 'informal waste handlers' refers to those individuals whose work involves waste management very generally. This can be picking, collecting, separating, transporting, processing or trading of domestic waste. They are not hired by the state and perform these activities independently from governmental programmes of waste management. Additionally, they tend to work in an economic structure that does not comply with legal, environmental or labour regulations. For these reasons, they belong to what officials think of as 'the informal economy'.

The operation and living conditions of waste handlers in Mexico City are well documented (Buenrostro and Bocco 2003; Castillo Berthier 2003; Medina 2005). Informal waste-pickers are called *pepenadores* in Mexican Spanish, from the indigenous Nahuatl language: 'to pick up'. Waste-pickers make a living by extracting valuable materials from domestic waste. These can be recyclable materials or objects that can be reused. For example, empty cosmetic bottles are sold to be refilled, rather than recycled. Food waste can also be selected, either to eat or to sell as animal feed.

Waste-pickers work at every stage of the waste management chain: in buildings, before the domestic waste is thrown away; on the street, where domestic waste accumulates; on the waste-collection truck; at every government facility (the Municipal Transfer Centres and the Waste Selection Plants) and on the landfills. Waste-pickers then sell the material they have gathered to recycling traders. Those recycling traders establish a bridge between waste-pickers and the formal recycling industry (Guibrunet *et al.* 2016).

Crucially, as the government does not have a formal programme of recycling (beyond one-off, neighbourhood recycling fairs), waste-pickers and informal waste handlers more generally handle the totality of the waste-separation process within the domestic waste and provide the necessary inputs to the formal recycling industry. The Inventory does represent waste-picking to the extent that it takes place in government infrastructure (transfer stations, waste selection plants and on the municipal trucks). In contrast, waste-picking, which occurs *before* it is even collected (that is to say, handled by *pepenadores* and households), does not exist as per the Inventory. Field research, involving interviews and ethnographic observations, reveals the existence and importance of informal waste workers in the city's recyclables value chains.

Conclusion

In recent years, urbanists have taken a critical perspective on urban metabolism by seeking to understand the extent to which urban metabolism is related to specific resource politics. Recent research has combined Industrial Ecology methods with the approach of Urban Political Ecology to consider issues of space (Cousins and Newell 2015) and resource distribution and governance (Baka 2014). Political-industrial ecology has been proposed as a sub-field that can produce research on both the material and social aspects of resource use (Newell *et al.* 2017); crucial to this endeavour is to mix methods emerging from the two fields, including quantitatively tracking material flows and qualitative research on the politics of such flows. At the heart of this attempt is the aspiration to appropriate urban metabolism methods to deliver environmentally and socially just cities that also inspire efforts towards a sustainability transition.

Our work has been deeply inspired by these attempts. However, we have grown concerned about what we see as the rise of urban-metabolism methods as a standard for environmental sustainability in urban areas. That rise corresponds to the growing need to control resources alongside the control of populations. Hodson and Marvin have popularised this concern as the concept of 'urban ecological security' that is transforming the approach to govern cities (Hodson and Marvin 2009). In this context, municipal workers and policymakers have turned towards urban metabolism as a means to gain efficiency in resource management, not as a means to deliver more progressive societies. This is our first attempt at thinking of urban metabolism as a governmental tool, one that is deployed with the objective of claiming authority in the city with the objective to control both populations and the natural resources that enable urban reproduction.

The case of Mexico City exemplifies this particular appropriation of urban metabolism: the waste inventory is a pioneering attempt to quantify waste flows. However, the framing of the tool responds to narrowly defined institutional priorities and needs, short-term rather than long-term objectives. The MFA focuses solely on the operations controlled by the government, and more specifically by the Ministry in charge of producing the Inventory. The Inventory produces a certain way of understanding and performing waste management, but does not reflect fully the complex webs of resources and actors that make waste management possible.

In particular, the MFA overlooks the processes of waste management that are thought of as being ungovernable. The delimitation of the administrative boundaries of the MFA, excluding outside flows and framing the city as a homogenous unit, has direct consequences for the city. On the one hand, it overlooks political debates and conflicts around waste that extend beyond the city limits. On the other hand, it ignores that waste management is affected by large inequalities in terms of service provision within the city.

Simultaneously, the MFA overlooks the role of different actors in waste management. Produced by the State, it focuses on the State's functions in waste management at the expense of recognising the importance of the private and informal sector in waste management. The former is important because it points towards the growth of a liberalised market of waste (as seen in other Latin American cities such as Santiago). The latter is important because of the need to recognise the role of informal workers and draw attention to the appalling health and working conditions in which they contribute to the crucial service of solid waste collection and management.

Urban metabolism points towards the politics of visibility in waste management. In the case of informal waste handlers, a lack of recognition in policy documents means a lack of recognition of the overall work they perform. This makes them invisible to other actors in the city. It prevents them from having a voice in decision-making processes. It also eludes the question of working conditions, workers' rights and workers' health. This is why researchers and activists have argued that a sustainable waste management system has to start by recognising the role of informal waste handlers (Scheinberg and Anschütz 2006; Wilson *et al.* 2009; Sembiring and Nitivattananon 2010; UN-Habitat 2010; Scheinberg *et al.* 2016; WIEGO n.d.).

Overall, our analysis of urban metabolism as a calculative practice points towards a core dilemma for research on material flows in cities. Are we looking to improve the methodology of urban metabolism to understand the inherent politics associated with it, by highlighting blind spots that could be improved? This would entail a revised approach that would enable a multi-scalar and spatially aware urban metabolism; or a multi-perspective approach. This is, for example, the aspiration of scholars involved in developing approaches for politico-industrial ecology, a debate that we have tackled head-on (Guibrunet *et al.* 2016). Rather, we adopt a critical perspective on urban metabolism as a calculative practice to render the urban fabric governable. Looking at it with a Foucauldian lens also suggests that we have to be wary of attempts to develop new mechanisms of control, and whether any transition – just or not – is already framed within a particular will to improve (Li 2007) whereby specific visions of the future are imposed over others. In this first attempt, we seek to reformulate our own ideas and point towards the underlying challenges faced by critical scholars of urban metabolism.

Notes

1 The waste inventory is available for download on the following page, which presents key data on waste management for Mexico City residents: www.sedema.cdmx.gob.mx/programas/programa/residuos-solidos.

2 This document will be referred to as 'Inventory' in the rest of the chapter. The full version of the Waste Inventory is publicly available at the following link as of 30 January 2017 (the waste-flows diagram appears on p. 75): www.sedema.cdmx.gob.mx/storage/app/media/programas/residuos-solidos/inventario-residuos-solidos-2014/IRS-2014.pdf.

References

Alfonso Piña, W.H. and Pardo Martínez, C. I., 2013. *Urban material flow analysis: An approach for Bogotá*. Colombia: Ecological Indicators.

Allen, A. and You, N., 2002. *Sustainable urbanisation: Bridging the green and brown agendas*. London: University College.

Baccini, P. and Brunner, P., 2012. *Metabolism of the anthroposphere: Analysis, evaluation, design*, 2nd ed. Cambridge, MA: MIT Press. Retrieved from http://ieeexplore.ieee.org/servlet/opac?bknumber=6267542.

Baka, J., 2014. What wastelands? A critique of biofuel policy discourse in South India. *Geoforum*, 54, 315–323. http://doi.org/10.1016/j.geoforum.2013.08.007.

Barles, S., 2007. *Mesurer la performance écologique des villes et des territoires : Le métabolisme de Paris et de l'Île-de-France*. Paris: Laboratoire Théorie des Mutations Urbaines.

Barles, S. 2009. Urban metabolism of Paris and its region. *Journal of Industrial Ecology*, 13 (6), 898–913.

Bristow, D.N. and Kennedy, C., 2013. Urban metabolism and the energy stored in cities. *Journal of Industrial Ecology*, 17 (5), 656–667.

Buenrostro, O. and Bocco, G., 2003. Solid waste management in municipalities in Mexico: goals and perspectives. *Resources, Conservation and Recycling*, 39 (3), 251–263.

Castán Broto, V., Allen, A. and Rapoport, E., 2012. Interdisciplinary perspectives on urban metabolism. *Journal of Industrial Ecology*, 16 (6), 851–861.

Castillo Berthier, H., 2003. Garbage, work and society. *Resources, Conservation and Recycling*, 39 (3), 193–210.

Chertow, M., 2008. 'Uncovering' industrial symbiosis. *Journal of Industrial Ecology*, 11 (1), 11–30.

Chrysoulakis, N., Lopes, M., San José, R., Grimmond, C.S.B., Jones, M., Magliulo, V. ... Cartalis, C., 2013. Sustainable urban metabolism as a link between bio-physical sciences and urban planning: the BRIDGE project. *Landscape and Urban Planning*, 112, 100–117.

CONAPO, 2010. *Índices de marginación* [*Marginalisation indices*]. Retrieved 23 January 2017, from www.conapo.gob.mx/es/CONAPO/Indices_de_Marginacion_Publicaciones.

Cousins, J.J. and Newell, J.P., 2015. A political–industrial ecology of water supply infrastructure for Los Angeles. *Geoforum*, 58, 38–50.

Currie, P., Lay-sleeper, E., Fernández, J.E., Kim, J. and Musango, J.K., 2015. Towards urban resource flow estimates in data scarce environments: the case of African cities. *Journal of Environmental Protection*, 6 (September), 1066–1083.

Dean, M., 2009. *Governmentality: Power and rule in modern society*. London: Sage.

Delgado Ramos, G.C., Campos Chávez, C. and Rentería Juárez, P., 2012. Cambio Climático y el Metabolismo Urbano de las Megaurbes LatinoAmericanas [Climate change and the urban metabolism of Latin American megacities]. *Hábitat Sustentable*, 2 (1), 2–25. Retrieved from http://remaderas.ubiobio.cl/ojs/index.php/arquitec/article/viewArticle/378.

Desrochers, P., 2002. Cities and industrial symbiosis – some historial perspectives and policy implications. *Journal of Industrial Ecology*, 5 (4), 29–44.

Ferrão, P. and Fernández, J., 2013. *Sustainable urban metabolism*. Cambridge, MA: MIT Press.

Fischer-Kowalski, M., 1998. Society's metabolism – the intellectual history of materials flow analysis, Part I, I 860–I 970. *Journal of Industrial Ecology*, 2 (1), 61–78.

Florisbela dos Santos, A.L. and Wehenpohl, G., 2001. El sector informal y los residuos sólidos municipales en México y Brasil [The informal sector and municipal solid waste in Mexico and Brasil]. *Gaceta Ecologica*, 60, 70–80.

Gabriel, N., 2014. Urban political ecology: environmental imaginary, governance, and the non-human. *Geography Compass* 8 (1): 38–48.

Girardet, H., 2004. *Cities people planet*. Chichester: John Wiley and Sons.

Gobierno del Distrito Federal, 2010. *Programa de Manejo Integral de Residuos Solidos del Distrito Federal* [*Programme for the integral management of the Federal District's solid waste*]. Mexico City: Gaceta Oficial del Distrito Federal.

Guibrunet, L., Sanzana Calvet, M. and Castán Broto, V., 2016. Flows, system boundaries and the politics of urban metabolism: waste management in Mexico City and Santiago de Chile. *Geoforum*, 85.

Hammer, M., Giljum, S., Bargigli, S. and Hinterberger, F., 2003. *Material flow analysis on the regional level: Questions, problems, solutions* (No. 2). NEDS Working Paper, Hamburg.

Hartmann, C.D. 2012. Uneven urban spaces: accessing trash in Managua, Nicaragua. *Journal of Latin American Geography*, 11 (1), 143–163.

Heynen, N.C., Kaika, M. and Swyngedouw, E., 2006. *In the nature of cities: Urban political ecology and the politics of urban metabolism*. London: Taylor & Francis.

Hodson, M. and Marvin, S., 2009. 'Urban ecological security': a new urban paradigm? *International Journal of Urban and Regional Research*, 33 (1), 193–215.

Hoekman, P., 2015. *Urban-scale material flow analysis in a South African context*. Cape Town: University of Cape Town.

Hoornweg, D. and Bhada-Tata, P., 2012. *What a waste – A global review of solid waste management*. Washington, DC: World Bank. Retrieved from http://scholar.google.com/scholar?hl=en&btnG=Search&q=intitle:WHAT+A+WASTE+A+Global+Review+of+Solid+Waste+Management#3.

INEGI, 2010. Censo de Población y Vivienda 2010 [Population and Housing Census, 2010] accessible online at inegi.gob.mx.

Kaika, M., 2005. *City of flows: Modernity, nature, and the city*. New York: Routledge.

Kennedy, C., Cuddihy, J. and Engel-Yan, J., 2007. The changing metabolism of cities. *Journal of Industrial Ecology*, 11 (2), 43–59. Retrieved from http://onlinelibrary.wiley.com/doi/10.1162/jie.2007.1107/abstract.

Kennedy, C., Pincetl, S. and Bunje, P., 2011. The study of urban metabolism and its applications to urban planning and design. *Environmental Pollution*, 159 (8–9), 1965–1973.

Kennedy, C.A., Stewart, I., Facchini, A., Cersosimo, I., Mele, R., Chen, B. … Sahin, A.D., 2015. Energy and material flows of megacities. *Proceedings of the National Academy of Sciences of the United States of America*, 112 (19), 5985–5990. http://doi.org/10.1073/pnas.1504315112.

Li, T.M., 2007. *The will to improve: Governmentality, development and the practice of politics*. Durham, NC: Duke University Press.

Linzner, R. and Ali, M., 2013. Briefing: waste management sustainability in low-income countries. *Proceedings of the ICE – Waste and Resource Management*, 166, 49–51.

McFarlane, C., 2013. Metabolic inequalities in Mumbai: beyond telescopic urbanism. *City*, 17 (4), 498–503.

Medina, M., 2000. Scavenger cooperatives in Asia and Latin America. *Resources, Conservation and Recycling*, 31 (1), 51–69.

Medina, M., 2005. Serving the unserved: informal refuse collection in Mexico. *Waste Management & Research : The Journal of the International Solid Wastes and Public Cleansing Association*, ISWA, 23, 390–397.

Moore, J., Kissinger, M. and Rees, W.E., 2013. An urban metabolism and ecological footprint assessment of Metro Vancouver. *Journal of Environmental Management*, 124, 51–61.

Monstadt, J., 2009. Conceptualizing the political ecology of urban infrastructures: insights from technology and urban studies. *Environment and Planning* A, 41 (8), 1924–1942.

Newell, J.P. and Cousins, J.J., 2014. The boundaries of urban metabolism: towards a political – industrial ecology. *Progress in Human Geography*, 5 December. https://doi.org/10.1177/0309132514558442.

Newell, J.P., Cousins, J.J. and Baka, J., 2017. Political-industrial ecology: an introduction. *Geoforum*, 85, 319–323.

Rice, J.L., 2010. Climate, carbon, and territory: greenhouse gas mitigation in Seattle, Washington. *Annals of the Association of American Geographers* 100 (4): 929–937.

Rosado, L., Niza, S. and Ferrão, P., 2014. A material flow accounting case study of the Lisbon metropolitan area using the urban metabolism analyst model. *Journal of Industrial Ecology*, 18 (1), 84–101.

Royden Turner, S., 2012. *From eradication to intervention – Urban informal ecosystem*. Cape Town: University of Cape Town.

Salinas Cesáreo, J., 2016. Basura de la CDMX, causa de la contaminación: Eruviel [Mexico City's waste, the cause of pollution, says Eruviel]. *La Jornada*, 16 March. Retrieved from www.jornada.unam.mx/ultimas/2016/03/16/basura-de-la-cd-eruviel-8616.html.

Schandl, H., Fischer-Kowalski, M., West, J., Giljum, S., Dittrich, M., Eisenmenger, N., Geschke, A., Lieber, M., Wieland, H. and Schaffartzik, A., 2016. *Global material flows and resource productivity*. Assessment Report for the UNEP International Resource Panel, UNEP.

Scheinberg, A. and Anschütz, J., 2006. Slim pickin's: supporting waste pickers in the ecological modernization of urban waste management systems. *International Journal of Technology Management and Sustainable Development*, 5 (3), 257–270.

Scheinberg, A., Nesic, J., Savain, R., Luppi, P., Sinnott, P., Petean, F. and Pop, F., 2016. From collision to collaboration – integrating informal recyclers and re-use operators in Europe: a review. *Waste Management & Research*, 34 (9), 820–839.

Scheinberg, A., Spies, S., Simpson, M.H. and Mol, A.P.J., 2011. Assessing urban recycling in low- and middle-income countries: building on modernised mixtures. *Habitat International*, 35 (2), 188–198.

SEDEMA, 2015. *Inventario de residuos solidos CDMX-2014* [Solid Waste Inventory, Mexico City 2014]. Mexico City.

SEDEMA, 2016. *Programa de Gestión Integral de los Residuos Sólidos para la Ciudad de México 2016–2020* [Programme for the Integral Management of Solid Waste for Mexico City]. Mexico City.

SEDESOL, CONAPO and INEGI, 2012. *Delimitación de las Zonas Metropolitanas de México 2010* [Delimitation of Mexico's Metropolitan Areas 2010]. Mexico City.

Sembiring, E. and Nitivattananon, V., 2010. Sustainable solid waste management toward an inclusive society: integration of the informal sector. *Resources, Conservation and Recycling*, 54 (11), 802–809.

Smit, S., Kaviti Musango, J., Kovacic, Z. and Brent, A., 2017. Conceptualising slum in an urban African context. *Cities*, 62, 107–119.

Swilling, M. and Annecke, E., 2012. *Just transitions – Explorations of sustainability in an unfair world*. Claremont, South Africa: UCT Press.

Swilling, M., Robinson, B., Hodson, M. and Marvin, S., 2011. *Cities, decoupling and urban development: Scoping the challenges*. Project Report: University of Salford.

Swyngedouw, E.A., 1995. The contradictions of urban water provision: a study of Guayaquil, Ecuador. *Third World Planning Review*, 17 (4), 387.

UN-Habitat, 2010. *Solid waste management in the world's cities*. London: Earthscan.

Robinson, B., Musango, J., Swilling, M., Joss, S. and Mentz Lagrange, S., 2013. *Urban metabolism assessment tools for resource efficient urban infrastructure*. Paris: UNEP.

While, A., Jonas, A.E. and Gibbs, D., 2010. From sustainable development to carbon control: eco-state restructuring and the politics of urban and regional development. *Transactions of the Institute of British Geographers*, 35 (1), 76–93.

WIEGO, n.d. *The urban informal workforce: Waste pickers/recyclers* (Informal Economy Monitoring Study). Manchester: WIEGO.

Wilson, D., Araba, A., Chinwah, K. and Cheeseman, C., 2009. Building recycling rates through the informal sector. *Waste Management*, 29 (2), 629–635.

Wilson, D.C., Velis, C. and Cheeseman, C., 2006. Role of informal sector recycling in waste management in developing countries. *Habitat International*, 30 (4), 797–808.

Wolman, A., 1965. The metabolism of cities. *Scientific American*, 213, 179–90. Retrieved from www.pubmedcentral.nih.gov/articlerender.fcgi?artid=1256631&tool=pmcentrez&rendertype=abstract.

Zhang, Y., 2013. Urban metabolism: a review of research methodologies. *Environmental Pollution*, 178, 463–473.

8 Exploring the epistemic politics of urban niche experiments

Matthew Cook,[1] *Ralph Horne,*[2] *Stephen Potter*[3] *and Alan-Miguel Valdez*[4]

Introduction

Cities play a significant role in responding to climate change (UN 2014; IPCC 2014). A growing range of actors have sought to mobilise action in response to this urgent agenda, including transnational organisations, national governments and local administrations. Such interventions have fruitfully been considered in terms of experiments to recognise their often-tentative nature, the sense of testing or establishing practice and the ways in which they are used as means of supporting or contesting knowledge claims and discursive positions. Socio-technical experiments addressing the urban dimensions of climate change, often framed as 'smart city' programmes, create new forms of political space within the city, intersecting public and private authority. Experiments are primarily enacted through forms of technical intervention in infrastructure networks, drawing attention to the importance of such sites in urban climate politics (Castán Broto and Bulkeley 2013).

Analyses of urban climate governance are needed that engage with the multiple actors through which governing is conducted. As local authorities seek to engage with climate change with limited resources and knowledge they invariably turn to an enabling mode of governance that depends on engaging with multiple actors, discrete sources of financial assistance, and often on 'reframing' climate change as an issue related to core agendas (economic growth, congestion, air pollution, urban planning and so on). The patchwork of responses thus produced has been criticised as indicative of a lack of capacity to coordinate and deliver an integrated, planned approach to urban climate governance (Corfee-Morlot *et al.* 2011). Other authors, however, see it as a process of variation and selection that can usefully adapt to uncertainties associated with the feasibility of different approaches and difficulties in identifying the intended and unintended effects of governance actions and strategies (Rotmans and Loorbach 2010; Avelino and Grin 2017).

The relatively protected spaces in which innovation and experimentation take place, conceptualised as niches, have been presented as critical to the process of socio-technical change (Hoogma *et al.* 2004; Smith and Raven 2012). As innovative technologies are applied in a social setting, novel forms of social organisation co-evolve with technological artefacts and practices. Although the changes that result from urban niche experiments are often presented as logical, technical and rational, the actors and mechanisms that shape decisions are far from obvious, involving cultures, power relations and multiple logics that are profoundly political (cf. Chapters 1 and 2). The processes of sense-making within the niche may reshape political rationalities as well as be reshaped by them. Drawing on the work of Flyvbjerg (2001) on *phronetic planning*, this chapter contributes to the growing body of knowledge that explicates the epistemological politics of urban experiments.

We start in the second section by discussing the phronetic approach and growing interest in the issue of power and empowerment in innovation experiments. Case-study research is presented in the third section to provide in-depth insights on an urban niche experiment in which a smart innovation 'MotionMap' is introduced. MotionMap is a mobile phone application that draws data from a variety of sources including governmental and corporate databases as well as feeds from custom-made sensors, providing real-time data about travel flows in the city. Like its commercial counterparts, such as Google Maps or Waze, MotionMap provides multi-modal routing and congestion warnings. However, MotionMap is intended to also contribute to public-policy objectives, providing a platform for exploring a variety of approaches to mobility in the city.

The case study is analysed to identify the qualities of MotionMap as the visible, interactive interface for a heterogeneous array of infrastructures, databases, sensors, commuters, policymakers, visions and narratives assembled within a niche. Flyvbjerg's approach enabled us to explore how niche activities are shaped by political rationalities and, in turn, shape these. This chapter therefore considers how a smart transport application (re)configures who gets what and with what consequences. And more generally, we explore the mechanisms of power at play as the interests, identities and interventions are influenced and shaped by the knowledge configuration performed by the application. Crucially as part of this analysis, the epistemic controversies (e.g. in determining the relevant qualities of transport flow to include) involved in the making of the application are explored. Cumulatively, the analysis thus covers the four concerns that were described in Chapter 1 as being central to an epistemic-politics perspective, but this is achieved through the application of the phronetic approach.

The phronetic perspective

Studies of socio-technical transitions have only recently started to engage with spatial and urban disciplinary perspectives (Coenen and Truffer 2012; Truffer *et al.* 2015; Loorbach *et al.* 2016; Moore *et al.* 2017). In tandem, in response to challenges from multiple quarters for socio-technical studies to account for power (Geels 2014) there is a growing interest in the study of shifting power relations between actors in sustainability transitions (Avelino 2009; Avelino and Wittmayer 2016). Accepting that things evolve in unpredictable ways and alignments of multiple actors, technologies, knowledge, social structures and policy settings occur over particular periods of time and in particular urban settings, the essential starting observation is that it matters who does the aligning, how, where, when and in whose interests. Avelino (2009) reveals the paradoxical nature of empowerment within urban experiments, which often require participants in participative empowering processes to be already empowered with conventions, language, concepts and relationships. Additionally, strategic language games enable powerful actors to exercise control over supposedly open engagement processes through subtle shifts in the definitions of 'sustainability' or 'empowerment' (Avelino and Wittmayer 2016) which may take experiments in radically different directions.

Avelino's perspectives on empowerment can be usefully complemented with the insights about values, knowledges and power found in the work of Bent Flyvbjerg (1998, 2001, 2004) who proposed a *phronetic* approach to engage with power relationships and rationalities in planning. The Aalborg case shows how two powerful actors (the City Planning Department and Chamber of Commerce) draw on the same evidence base to make two quite different arguments that ultimately affect the Aalborg redevelopment project. The rational-choice discourse rests on the notion that if decisions are rational, then participation isn't required. The phronetic perspective, however, reveals how rationalities are not as value-neutral as they may appear to be. Focused on the *Real Politik*[5] of such projects, Flyvbjerg's work therefore questions the democratic focus on rationality. In citing Flyvbjerg's work, Avelino and Grin (2017) argue that (a) the work of analysing power rationalities and (b) that of making judgements about what *should* be, constitute the 'is' and 'should' approaches respectively in social sciences and both are valuable in socio-technical transitions studies. In other words, showing that multiple arguments may be made by drawing on the same sources reveals that rationality is far from value free and that research should attend to not only instrumental rationality, but value rationality too. In this way, Flyvbjerg turns the Baconian dictum 'Knowledge is Power' on its head – Power *is* knowledge. Attending to the 'should', he also encourages engagement with everyday practices and the need to generate useful results that *matter*.

Flyvbjerg (2001) draws upon the Aristotelian idea of *phronesis*, which is variously translated today as practical wisdom, practical judgement, common sense, or prudence as an overarching banner for his approach. Similar to other practice-based research (cf. Guy and Moore 2004), a phronetic approach amounts to a call for academics to engage with practice and help decision-makers critically reflect on, among other things, their own processes.

The principal objective of the phronetic approach is to clarify values, interests and power relations as a basis for praxis. It is argued by Flyvbjerg (2001) that this objective can be met by focusing on the following value-rational questions:

1 Where are we going?
2 Who gains and who loses, and by which mechanisms of power?

3 Is the development desirable?
4 What, if anything, should we do about it?

Flyvbjerg's questions are profoundly about power. Further, following the idea of phronesis, the 'we' referred to in questions one and four are researchers asking questions and those who share the concerns of the researchers in the community being studied. Here we should note that Flyvbjerg is not looking for universal rules, as the particularities of various contexts and phenomena must be taken into account. In other words, 'we' will always be situated in relation to a specific context. The absence of 'they' suggests that research is not conducted from 'no where' and that the results of research rightly affect developments. Here the approach stresses that no single actor has sufficient wisdom and experience to give complete answers to all four questions, whatever those questions might be for a specific problematic.

Phronetic research is problem-relevant research, which questions, among other things, how towns and cities are redeveloped. It is not method driven; indeed, various methods can be drawn upon as necessary to address specific issues at hand. Research that effectively answers the four value-rational questions detailed above as a basis for action is valued more than methodological orthodoxy.

The questions at the heart of phronetic research balance instrumental rationality with value rationality and increase the capacity of researchers, planners and those planned to think and act in value-relational terms. Such an approach inevitably leads to questions about foundationalism versus relativism: the view that there are central values that can be rationally and universally grounded, versus the view that one set of values is as good as another (Flyvbjerg 2001). To address this concern, Flyvbjerg argues that the rejection of foundationalism implied by *phronetic* research does not leave researchers 'normless'. Instead, the point of departure is their *attitude* to the situation in the context under study – informed by a common view among a specific reference group to which the researchers belong.

Thus, in researching the smart-transport experiment that took place in Milton Keynes (MK) we must be explicit about our own position, conscious of social difference and somewhat distrustful of neo-liberal narratives that often imply that efficiency and economic growth are key to the improvement of urban sustainability and quality of life. We oppose such a neo-liberal smart-city narrative because we question its epistemic grounds, particularly its logical positivism, which implicitly holds that the world is, in principle, perfectly knowable, its contents enumerable and their relations capable of being meaningfully encoded in the state of a technical system (Greenfield 2013). We also question the assumption that solutions to urban problems can be arrived at algorithmically and applied transparently, dispassionately and in a manner free from politics (ibid.). Particularly, as the authorship of an algorithm intended to guide the distribution of civic resources is itself an inherently political act (Kitchin 2017), we are concerned by narratives that make urban experiments apolitical, bypassing mechanisms of democratic accountability and rendering the mechanisms of power invisible. Thus, making such mechanisms of power visible is a central concern driving this research.

It cannot be overstated that power is central to phronetic research. Indeed, it argues that rationality without power is irrelevant (Flyvbjerg 1998), which chimes with Friedman's (1998) 'inevitable question of power'. In phronetic research both the Weberian questions 'Who governs?' and 'What governmental rationalities are at work when those who govern govern?' (Foucault 1979) are mobilised to explore value rationality in specific contexts. In turn, these emphasise the epistemic dimension of power: 'Knowledge and power, truth and power, rationality and power are analytically inseparable from each other; power produces knowledge and knowledge produces power' (Flyvbjerg 2004, p. 293).

The phronetic perspective proved highly relevant for the case study. Epistemic contests regarding the definitions of empowerment, sustainability, smartness and even motion, and the various alternatives to operationalise them through the sensors and data infrastructures of a 'smart city' proved germane to understanding the relation of knowledge and power in the development of MotionMap.

Case study – MotionMap

> Aristotle explicitly identifies knowledge of 'particular circumstances' as the main ingredient of phronesis. Foucault similarly worked according to the dictum 'never lose sight of reference to a concrete example' Phronetic research thus benefits from focusing on case studies, precedents and exemplars.
>
> (Flyvbjerg 2004, p. 135)

Smart technologies are often discussed in the abstract, with critics and advocates dissecting the concept itself with limited reference to actually existing smart cities (Shelton *et al.* 2015). When concrete examples are discussed, they are often documented from a technical perspective: academics working on developing smart-city technologies and policy formulations – especially those in the sciences and computational social sciences – position their work as pragmatic and non-ideological. In consequence, discussion of the politics, power and epistemic conflicts that take place as smart-city projects develop on the ground are frequently underdeveloped (Kitchin 2014). Here, the importance of unpacking the epistemological politics of urban experiments will be illustrated through case-study research that provides in-depth insights on a 'smart' urban niche experiment. A coalition of policy, technology and academy actors supported the development of a smart innovation, 'MotionMap', broadly envisioned as an empowering application that would use real-time data to transform the transport system of MK and contribute to the economic development of the city.

Background

MK is a rapidly growing new town that was founded in 1967. Prior to the designation of MK, the area had a population of less than 50,000. Today, it is estimated to have over 250,000 residents. Between 2004 and 2013, MK was the fastest growing UK city, expanding by 16.5 per cent (Centre for Cities 2015). Local government, universities and industrial actors found that the rapidly growing new town provided a suitable venue to explore the benefits of sustainable technologies and to develop new business models for them. Projects such as the demonstration of the world's first solar-powered house in 1972, pioneering energy standards in buildings in 1979 and the UK's first kerbside recycling collection scheme in 1992 contributed to the reputation of MK as an innovation test bed (PRP Architects 2010). The strategic positioning of MK as an experimental city has been deliberate, with the city council consistently developing the MK 'brand' as a test bed for sustainable living initiatives: a place where business and governmental actors can test new ideas and set standards for future adoption around the UK (PRP Architects 2010; MKC 2010; MKC 2011; MKC 2013).

Several urban transition experiments being deployed in MK are related to ideas of sustainable transport, a growing concern for city authorities. Despite its relatively small (but rapidly growing) population of approximately 250,000 people, the automobility problem in MK is eminently suburban. The city was built to a design dating from 1969–1971 that sought to facilitate the use of the private car for all journey purposes (Llewellyn-Davies *et al.* 1968): a one-kilometre grid of high-capacity, high-speed 'grid roads' and extensive car-parking in all areas, coupled with low-density development. MK has an extensive network of cycling and pedestrian paths (colloquially known as 'redways') but they were designed for leisure (Clapson 2013, pp. 15, 16, 64) and are widely considered unsuitable for commuting (Franklin 1999; Edwards 2001). MK is consistently ranked last or near-last in car dependency scorecards (i.e. amongst the most dependent) (Campaign for Better Transport 2014) and rapid city growth is expected to aggravate the issue. Because of the combination of high car dependence and rapid population growth, traffic growth of 60 per cent is expected in MK by 2026 (MKC 2012a), but local authorities will only be able to provide an extra 25 per cent capacity through junction improvements and other measures.

The city council has explored various informational and behavioural approaches for making more efficient use of the existing infrastructure through a variety of pilots centred on water, energy and transport technologies, among others. In the case of MotionMap, the interests of the council, technology developers and academia were aligned, with a smart-city project providing an arena for exploring the application of real-time data for addressing transport problems.

Milton Keynes Council has traditionally supported transition experiments through an approach based on enabling governance: instead of funding and controlling traditional solutions, the council tends to encourage and support various commercial, academic and community actors interested in running niche experiments. MotionMap, the application analysed in this case study, was not commissioned by the council, but was part of the council-facilitated MK:Smart, a £16 million (circa 18.25 million euro) smart-city programme involving a network of industry, community and research actors. The programme, funded by the Higher Education Funding Council for England (HEFCE), was led by the Open University. MotionMap was the key transport output of the programme, based on the premise that high availability of personalised information on travel choices could have a similar, if not greater, impact to that of Personalised Travel Plans which have been used to provide enhanced travel information for households. Traffic reductions of up to 11 per cent have been achieved through the adoption of such plans (Cairns *et al.* 2008).

Design and development of the application was led by the University of Cambridge, in collaboration with the Open University and various start-up companies with the technical skills required for deploying sensors and developing visualisations, analytics and routing algorithms. The project was also supported by Milton Keynes Council and by partners such as BT, Huawei, Samsung and Tech Mahindra, who contributed to the data infrastructures for the smart-city project in which MotionMap was embedded, but did not have direct input in the design of the application. Data from a variety of sensors and feeds relevant to transport, water and energy were stored and made available through a 'Data Hub' managed by the OU as part of the MK:Smart programme (d'Aquin *et al.* 2014). Data were made publicly available with the expectation that SMEs, entrepreneurs and civic hackers would develop their own smart applications. A Cambridge-based start-up was commissioned by MK:Smart to develop MotionMap as a flagship application for the transport work package.

Project leaders envisioned MotionMap as an application accessible through mobile devices that would allow users to make spontaneous public transport decisions with increased efficiency and convenience. The real-time sensing and visualisation of travel flows made possible by MotionMap would facilitate 'spontaneous real-time choices about transport which spring from the exercise of personal preference rather than bureaucratic coercion' which were expected to lead to substantial (and beneficial) changes to the nature and pattern of movements within the city' (MK:Smart, 2014). Like its commercial counterparts, such as Google Maps, CityMapper or Waze, MotionMap was designed to provide multi-modal routing and congestion warnings. However, MotionMap was also intended to contribute to public policy objectives, providing a platform for exploring a variety of approaches to mobility in the city. Real-time data about transport flows captured through the application and its related sensing infrastructure would feed into city dashboards and inform the city's traffic models.

The somewhat abstract concept of 'smart transport', and particularly the unexpectedly malleable ideas of 'transport flows' and 'motion', were subject to exploration, negotiation and epistemic debate throughout the life of the project.

Method

A variety of qualitative data from primary and secondary sources were collated and analysed following a clustering and coding method (cf. Miles and Huberman 1994). Policy and corporate literature on smart cities provided insight about the stated goals of national and transnational actors in policy and technology organisations. Sources of data included:

- Promotional materials and articles in popular media (Cisco 2014; Sheffield Smart Lab 2015; DriveMidlands 2016; Connect Cambridgeshire 2016; Cambridge News 2016; OneMK 2016);
- A smart-city perspective study by MK Council (MKC 2012b);
- A feasibility study assessing MK's potential participation in the future cities demonstrator programme (TSB 2013);
- A comparative analysis of the feasibility studies performed by a transnational consultancy firm (TSB and ARUP 2013);
- A long-term vision for the city (MKF2050 2016), and;
- Academic articles produced by researchers working in MK:Smart (d'Aquin *et al.* 2014; d'Aquin *et al.* 2015; Potter *et al.* 2015; Wolff *et al.* 2015a; Wolff *et al.* 2015b; Montaner *et al.* 2015; Gooch *et al.* 2015; Okada *et al.* 2015; Gaved and Peasgood 2015; Williamson 2015; Caird *et al.* 2016; Valdez *et al.* 2018).

The somewhat idealised accounts about the development of the smart city identified in secondary data were contrasted with primary data collected during a series of MK:Smart citizen engagement workshops and secondary data produced by members of the coalition responsible for the smart-city programme in MK.

Primary data were predominantly collected through the observation of a series of citizen engagement workshops through which OU researchers (including two of the authors of this chapter) sought to engage citizens as co-creators of MotionMap. Discussions prompted through iterative interactions of participants and prototypes were used to explore the co-constitution of the technical and the social. Citizen groups interested in specific transport issues were contacted through community coordinators with extensive links to the voluntary and community sectors (CAMK 2014a, 2014b). Given the relatively low levels of bus usership, walking and cycling for commuters in the city (4 per cent, 7 per cent and 3 per cent respectively), this was necessary so that the populations of interest could be reached. Four of the workshops were open to the general public but targeted citizen groups interested in specific transport issues, who were contacted through community coordinators; the remaining two workshops were closed and the participation of specific project partners was sought (MK:Smart partners and prospective entrepreneurs interested in developing data-driven business models). Twelve to forty participants attended each workshop and were separated into smaller groups to facilitate discussion. Workshop guides with suggested questions and topics to discuss were developed and used to stimulate group discussions. Data were collected through notes taken by workshop facilitators working with each group, from feedback forms provided to participants and from incidental outputs of the group exercises (for example, sticky notes and flipboard sheets with notes from brainstorming sessions). The workshops were used for data collection, but also to inform the design of the application and to interrogate the assumptions of the 'smart' experiment itself. The original vision for MotionMap, with real-time data being used to support a more efficient use of physical infrastructure, implied a particular conception about the relationship between power and urban knowledge: citizens would be 'empowered' to make more informed transport choices.

While this vision of empowerment was not interrogated critically in the early stages of the project, MK:Smart as a whole had a different (yet compatible) ethos of empowerment, one where citizens would be 'empowered' through their participation as co-creators of smart-city applications. The workshops were meant to support this conception of empowerment, which is widely held in academic and corporate literature on smart cities. Building on the premise that city authorities and planners may not always be able to predict which smart urban services are needed to tackle issues as they arise, the beneficial economic and societal outcomes of smart-city programmes are expected to emerge through a mix of formal planning, market forces and, crucially, citizen involvement (Greenfield 2013; Vilajosana *et al.* 2013; BSI 2014; Osborne Clarke 2015; Hollands 2015).

The workshops used non-functional mock-ups, intended to be very abstract and open to interpretation. For example, one of the very first mock-ups was a static image of a 'heat map' without any labels and the meaning of the heat map was left intentionally ambiguous so it could be interpreted, for example, as an indicator of traffic congestion or as pedestrian density. Those limited prototypes were refined in an iterative fashion, with the expectation that results from the workshop would feed into the specifications used by the software development team. Abstract mock-ups were followed by non-functional mock-ups in which the expected functionality was made more explicit (Figure 8.1). Later, interactive prototypes with limited functionality and a real connection to sensors

Figure 8.1 Artist representations by the development team illustrating concrete examples of potential applications for MotionMap.

Source: credit: MKSmart.org.

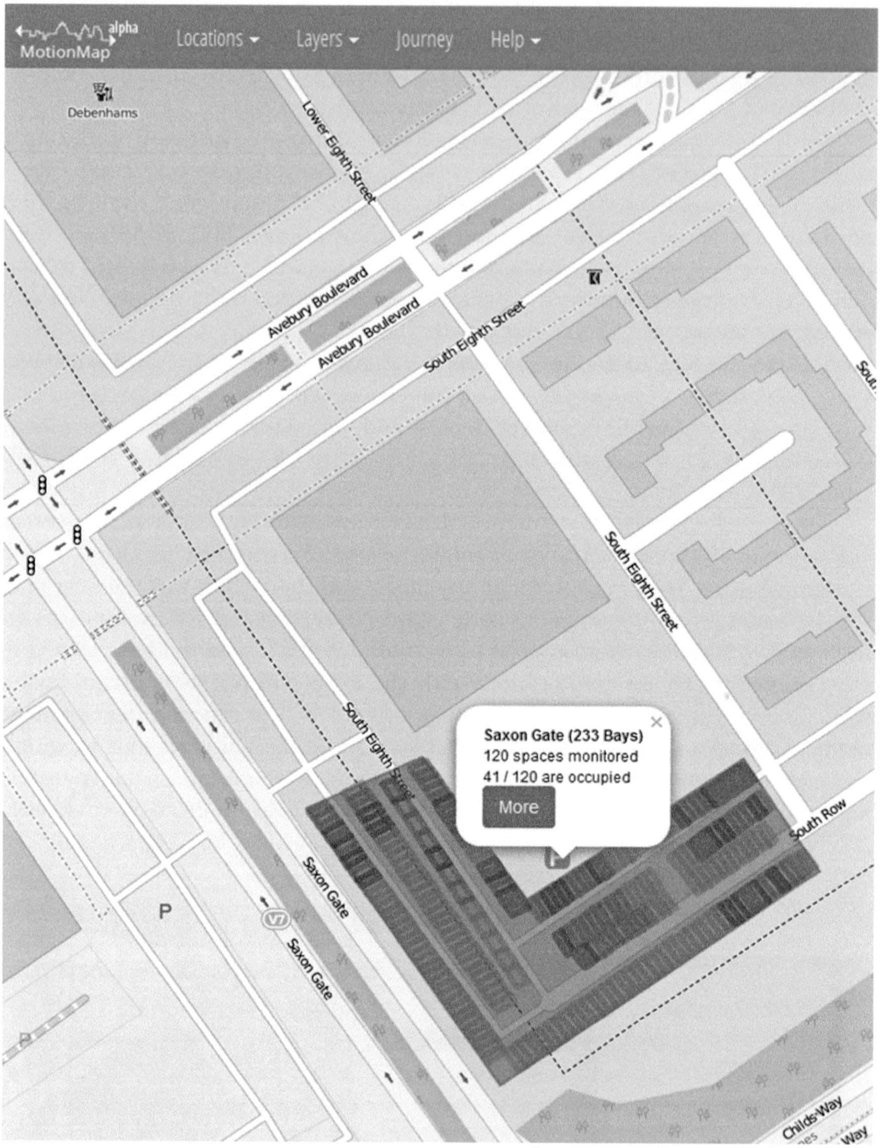

Figure 8.2 Screen capture of the actual functional MotionMap prototype, which relayed real time information for selected areas. In this case it shows which car parking spaces are occupied and which are free.

Source: credit: MKSmart.org.

deployed in the city centre became available and were included in the discussions (Figure 8.2).

Analysis

Where are we going? Who wins and who loses?

Smart-city projects are frequently defined in vague, optimistic language so that they may claim a broad legitimacy for guiding stakeholders and constitute an attractive reference for actors at all levels and across sectors (Wolfram 2012). However, it remains rather open what the actual pursuit of a 'smart city' is and, therefore, which winners and losers we are to expect from its deployment (ibid.). In the case of MotionMap, iterative encounters between citizens and prototypes revealed that there was not a unique answer to the phronetic questions. There was a tension, which remained unresolved until the conclusion of the project, between various visions about the potential direction of the experiment. The design of early prototypes was shaped by a strategy based on the use of big data to increase the efficiency of transport infrastructures: the sensing infrastructure of the smart city would be used to make citizen activities visible so that users could voluntarily manage their demand of transport services. Citizen workshops suggested a subtly different application, however, that implied big differences in visions about real-time knowledge and power.

Early MotionMap prototypes implemented the original vision of the project leadership and city council, which assumed that users would value transport information if they could use it to travel more efficiently, avoiding congestion by changing the route or time of their journeys, or avoiding travel altogether if efficient alternatives were not available (voluntary demand management). The application was largely intended to cater to drivers, as 71 per cent of the commuters in MK rely on private cars for their daily commute. However, workshop participants did not think that the system as designed would lead to significant changes in their car-driving behaviour. Habitual drivers were already familiar with the patterns of congestion in the city:

> We have been having this conversation since 30 years ago. Gridlock happens two hours out of 24 and everyone knows it, but everyone has to get to work at peak hour anyway.

In contrast to the relative scepticism of drivers, workshop participants that relied on public transport, walking and cycling were optimistic about the potential impact that real-time information would have on their travel experiences. However, they did not discuss the expected benefits in terms of increased efficiency. Participants generally considered that all alternatives to the car involved a loss of control and reliability. Non-drivers saw the potential to use smart technologies to have a more reliable experience, to increase their control over their travel and to increase the accountability of public-transport providers. Thus, in addition to being 'empowered' to make better choices, they also expected to be empowered to exert pressure for change in the transport system.

For example, many bus stops in MK have real-time information boards, but users found the information unreliable. Several users reported that sometimes they would spend 20–30 minutes waiting by the bus stop for a bus that according to the official information was just around the corner and sometimes buses would seem to just disappear from the application. Their rough estimate was that the information was accurate 90 per cent of the time and this became problematic for frequent bus users, particularly when planning multi-leg, multimodal journeys.

Bus users felt disempowered by the situation, stating that bus companies appear to dictate their own rules with little accountability to users or local authorities. Data related to scheduling, routing and incident reporting was distrusted, as there was a perception that 'it's very easy to manipulate data if you are a bus company'.

Workshop participants considered that multi-modal journeys would greatly benefit from sensing systems complemented through crowdsourced, real-time reports about transport services. This information would be of immediate use to other bus passengers and the accumulated records provided by this monitoring activity would make transport providers more accountable. This, in effect, would counteract the existing disempowering relationship between bus users and service providers.

Cyclists and pedestrians were also interested in crowdsourced information that would let them know about hazards like flooded underpasses, footpaths with insufficient night-time lighting, or cycle paths made dangerous because of broken glass. While the specific information required by bus users, cyclists and pedestrians was different, the intended use for information was similar in all cases: users wanted to have the information available in real time so they could re-route around broken links and they also wanted to know that information would be kept on record so city managers and service providers could be held accountable.

The original vision for MotionMap could potentially address the traffic problems caused by the rapid growth of the city, but it would largely maintain the status quo, with the same power structures in place and the same mix of transport modes but operating more efficiently. The alternative suggested by citizens would complement sensor data with crowdsourced reports to give users more power over transport service providers. Instead of making citizens (as commuters) visible to increase their efficiency, the actions of transport and infrastructure providers would be made visible, increasing accountability and potentially challenging existing power structures.

What mechanisms of power are in play?

Data were central to various mechanisms of power observed in the MotionMap experiment, either through direct control of its access or, more subtly, by controlling the categories and definitions used to understand urban flows. Smart cities make vast volumes of urban information available to citizens, businesses and city managers. Data is drawn from a wide variety of sources and systems, collected through pervasive and ubiquitous computing and digitally instrumented devices built into the very fabric of urban environments. An unprecedented level of access to data about the city is expected to facilitate the real-time management of its resources, improving the efficiency of public-service delivery, the sustainability of the urban environment and the quality of life of citizens (Cisco 2012; BIS 2013). This was, indeed, the approach pursued by the designers of MotionMap:

> We're focusing our efforts on trying to extract all the information that is available around the city and putting it together in a form that is easy to understand and use. So, we're developing an app called MotionMap. Its purpose is to be able to tell you on your handheld device, in a very simple way, how busy it is at any place in the city … Our job now is to research how best to do that. How do we measure busyness? How do we put this notion of busyness in the hands of the user? How then might you nudge the users to make the decision which will help the overall picture of congestion be relieved?
>
> (MK:Smart 2014)

Here, we use the term 'epistemic controversy' to characterise the tensions induced by disagreements about the best way to 'extract all the information that is available around the city'. The field of urban studies is subject to periodic epistemic crises. A recurrent question in the epistemology of the urban is: through what categories, methods and cartographies should urban life be understood (Brenner and Schmid 2015)? The latest of those crises has been caused by the rapid penetration of sensors and informational infrastructures capable of processing huge data assemblages (Gleeson 2014, p. 348). Deployment of such infrastructures is often pursued uncritically, based on the misleading assumption that 'data are transparent, that information is self-evident, the fundamental stuff of truth itself. If we are not careful, in other words, our zeal for more and more data can become a faith in their neutrality and autonomy, their objectivity' (Gitelman and Jackson 2013, pp. 2–3).

In the case of MotionMap, epistemic controversies centred on two issues: the selection of data sources and the co-design of the envisioned functionalities for the application.

In practice, only a fraction of the data sources sought by the developers of MotionMap could be negotiated and implemented. For example, it was not possible to reach a workable compromise when seeking permission for installing 'busyness' sensors in commercial areas: managers of commercial spaces considered that footfall information was commercially sensitive and were unwilling to make it publicly available, but keeping it private would have gone against the open data ethos of the MK:Smart project. Additionally, it became clear that the interests of users and those of business owners were not always aligned. Part of the vision for MotionMap was that a user might see the city centre being really busy and change their plans to come another day or choose to go to a less busy department store over another more crowded one. The managers of the shunned store would be unlikely to be pleased in that hypothetical scenario.

Similar concerns over the commercial sensitivity, cost effectiveness and technical feasibility of using various data sources led to a series of negotiations with other stakeholders such as taxi and bus companies and parking managers, some more successful than others. Consequently, data streams feeding into the early versions of MotionMap were limited compared to the original vision. Available feeds were, nevertheless, sufficient for the purposes of development and included interactive maps, real-time routing information provided by public transport operators; occupancy information from managers of a small number of parking areas in the city centre; and main road–traffic information purchased through 'data as a service' agreements with corporate providers.

While negotiating access to pre-existing data and to senseable spaces was important for defining the nature of MotionMap, data that were not available from existing sources could potentially be captured by deploying new sensors. A variety of sensing devices were deployed in certain areas of MK specifically for these purposes. Various approaches were trialled, but in the end automated image analysis of CCTV was selected because of its cost effectiveness and ease of deployment. The sensors, which relied on inexpensive cameras, microprocessors and sophisticated algorithms, were originally designed to help owners and managers of public spaces to understand exactly how those spaces are used, to facilitate a more intelligent and dynamic allocation of resources and to better forward plan (Cisco 2014). Drawing on lessons learned from the MotionMap project, the start-up company developing the technology updated their vision, emphasising how solutions based on their technology can empower citizens so they can make better choices about how they use their city (Sheffield Smart Lab 2016). Arguably, this is a somewhat limited conception of empowerment, as users are only 'empowered' to make more efficient choices within the existing system and structures of power. However, as discussed in the section above, *Where are we going? Who wins and who loses?*, real-time information such as the one provided by the sensors could potentially be used to facilitate other forms of empowerment more disruptive of the status quo.

Is the development desirable? What, if anything, should we do about it?

As discussed in the second section of this chapter, answers to phronetic questions cannot claim to be rational and value neutral but are grounded in the researchers' attitudes, which must be made explicit. Thus, we restate our commitment to equal, democratic and liveable cities and also, importantly, our concern with making the mechanisms of power visible. Urban experiments are often framed as 'empowering', with policymakers claiming that their initiatives will 'empower' 'the community'. Critical perspectives on empowerment, however, emphasise that attempts to empower others may have the paradoxical effect of disempowering them through the creation of a new dependency relation (e.g. Hardy and Leiba-O'Sullivan 1998). The spread of 'participative events' throughout a policy field may even have counter-productive effects as the 'participants' may be confirmed in their overall impression that they are 'pawns in some new policy trend' (Avelino and Wittmayer 2016). Avelino (2011) suggests that the success of empowerment in transitions can only be evaluated 'at the moment of physical materialisation, possession and profit': who has materialised the new institutional arrangements, who possesses the new resources, who profits from them and who's goals have been realised?

Reflecting on Avelino's questions, MotionMap had limited success as a means of empowering its users. It proved difficult for the MotionMap development team to pursue a model that would give users ownership and control of the data, or one that would realise the goals identified through the workshops. In development, the focus for MotionMap shifted from active user engagement and user ownership of data (e.g. through crowdsourcing data on bus, parking and cycleways) towards technical solutions to obtain the same data (visual analytic sensors). Various aspects of the development process contributed to this.

One aspect was that the project had a strong technical team and capabilities built around the gathering of data from instrumented sources. Its background and culture, although open to learning from co-creation and crowdsourcing experience, was not centred on such approaches. This is a common situation in big-data smart-city projects and also in commercial big-data approaches (such as the way Google Maps detects mobile phone signals to map traffic congestion).

A second aspect was that, even if the team had sought a crowdsourced rather than an instrumentation approach, user 'buy-in' to MotionMap was low and selective. Crowdsourcing could be viewed as inferior to sensors in providing reliable information. Even established commercial crowdsourcing systems have very patchy reporting around MK – and is at a level far too low to generate reliable data for an application like MotionMap. This suggests that smart-city projects seeking an empowering approach may need an initial phase built around instrumentation and sensors. This will provide the functionality to build up the critical user base to then move on to include enhancing by crowdsourced information. However, this sort of model very much constrains the co-creation role for users.

A third aspect is that users sought empowerment for a different purpose than was envisaged by the team developing MotionMap. The overarching aim of MotionMap was to provide functional travel information that would help users to make more-informed travel choices. That would then yield benefits in terms of congestion reduction and economic development. But in the workshops, this functionality was not particularly valued. It could be that, in practice (like many new technology products), people would find MotionMap more useful than envisaged hypothetically. This could well prove to be the case. However, it was notable that the workshop participants sought empowerment through MotionMap for very different purposes than travel information functionality. One was they saw it as a way for them to work as communities of interest (e.g. cyclists reporting to each other glass on the cycle paths). Customising for specialist interests was also mentioned (which requires the application to be designed in a way that can be customised with relatively basic programming knowledge through an Application Programming Interface (API)).

A rather different empowering purpose was that users wanted to use MotionMap to hold providers and authorities to account. For example, crowdsourced information might be used to ensure the reliability and punctuality of the bus service or to ensure that pedestrian areas have sufficient illumination and are in good repair. Through ownership and control of transport data, users would counteract the imbalance of power and their dependency on transport service providers. This was particularly strongly felt by bus users and cycle groups. However, there is a situation that, due to the existing power imbalance, such groups have a lobbying and pressure group culture and regime. They thus react within this regime and found it quite difficult to envisage how MotionMap might shift them to become co-creators of a service, rather than lobbying the service providers.

This set of purposes unrelated to travel functions did not readily map onto the original aims for MotionMap of congestion relief and economic development. In consequence, the meaning and the potential of 'Smart Transport' was contested. There is certainly potential for using smart technologies to enhance traditional city and transport planning aims, but such optimisation may not be deemed desirable by users. Once new technologies such as big-data applications are deployed to this domain, it should be recognised that the desires and aspirations of consumers in the digital economy will begin to emerge. People will no longer view themselves as passive users of functional services, but as interactive users that address wider lifestyle desires. The tension in the purpose of MotionMap is one of seeking to address traditional city-planning and transport aims with a smart application powered by big data, moving into the digital technology sphere. It perhaps needs to be recognised that this will generate new user expectations for interactive and co-creation involvement that will become as much a part of public policy as the traditional engineering and planning goals around which professional practice and competencies have been built. As transport policy initiatives move into the digital arena, analogue behaviours and thinking need to shift too.

Epilogue and discussion

At the time of writing, the MK:Smart programme reached its formal conclusion. The conditions of funding stipulated that the programme would run until 2017, and that projects initiated by the programme would become independent and financially self-sustainable afterwards. Having secured the support of local authorities in MK and Cambridge, the MotionMap application is still being developed by one of the start-up companies in the original coalition. A second start-up maintains and expands the sensing infrastructures deployed for the

MK:Smart experiment, with funding and support from the city council. The sensor network associated with the application is nearing the levels of coverage needed to deliver practical results, providing real-time information on approximately 70 per cent of the buses used by the main transport provider in the city, 85 per cent of the pedestrian activity in the city centre, nearly 100 per cent of the traffic in the main grid road network, and thousands of parking spaces. MotionMap as a concept remains attractive from a policy perspective. Through a series of planning activities taking place between 2016 and 2017, MotionMap was made part of the medium and long-term transport strategy advocated by the MK Futures 2050 Commission, an independent body launched by Milton Keynes Council to address potential longer-term futures for MK. MotionMap is envisioned as an enabler of new forms of mobility to ensure everyone who lives, works, studies or does business in the city is able to 'move freely and on-demand' (Transport Innovation Task and Finish group 2016; MKFC2050 2016). The sensors and data infrastructures developed for the niche in which MotionMap was initially developed are being adapted to other applications and locales (UKAuthority 2016; Highways Industry 2017) and MotionMap is travelling to new locations (Connect Cambridgeshire 2016; Cambridge News 2016), where it is being adapted and has become the subject of new epistemic debates. There is potential for the smart-transport innovation to break out beyond the experimental niche, leading to a transition towards a smart region, even as the meaning of smart transport in MK is still being explored and contested.

The case study reveals how various actors (including two universities, the local authority and citizens) engaged in the development of this application have further rendered the MK mobility socio-technical system an object of urban governance. Prototype MotionMap designs were developed by and presented to various actors in multiple institutional arena: the MK Transport Innovation group, Citizen Labs and project team meetings. This process focused on deciding which transport modes and infrastructures to map and the qualities of these that would be most relevant and most readily translated into real-time data. Qualitative data collected revealed a rich picture of how arguments are made to justify views of the mobility system, how powerful actors were mobilised through notions of 'efficiency' to effect transition to more sustainable urban mobilities' (e.g. to reduce carbon emissions) and how temporal, spatial, institutional and socio-cultural boundaries are constructed in urban experiments and shape such processes.

Here, we found that a phronetic perspective proved useful to inform the theoretical sensitivity of the research, even if definite answers could not be reached for a smart-transport application that is still being developed and negotiated. Additionally, we found that the task was made even more difficult because of language games that become part of the process of negotiation and a mechanism of power. Notions of sustainability and smartness became strategically malleable (Smith and Stirling 2008; Avelino and Grin 2017). For example, increased efficiency in the use of the road network may be primarily motivated by cost reductions but is framed as a mechanism for reducing congestion (i.e. good for 'people') and emissions (i.e. good for 'planet'). In this way, the primary goal of economic optimisation is framed in terms of 'sustainability' (Avelino and Grin 2017).

In retrospect, the development of MotionMap can be seen as an exercise in developing a local version of the idealised smart-city narrative and in balancing its inherent tensions. Various institutions use narratives to elevate some imagined futures above others, according them a dominant position for policy purposes, which draws attention away from alternatives (Van Hulst 2012; Jasanoff 2015). Coalitions of policy and industry actors engage in a collaborative storytelling through the publication of advertising materials, technical prospects, frameworks and policy documents, often conceptualised as roadmaps. This smart-city narrative provides direction to the imaginaries and practices of actors concretely building cities through particular pilot projects (Bakıcı *et al.* 2013; Söderström *et al.* 2014; Lee *et al.* 2014). Those idealised narratives make compelling promises of increased efficiency, economic growth, sustainability, quality of life and citizen empowerment (Shapiro 2006; Hollands 2008; Aoun 2013; BIS 2013; Kitchin 2014; Neirotti *et al.* 2014; Albino *et al.* 2015). The various goals in this vision are generally presented as interconnected and it is implied that they will emerge jointly following the deployment of smart infrastructures. In the case of MotionMap, however, the various promises made by the smart-city narrative proved to be somewhat independent from each other. Hence the relevance of the phronetic perspective, that may usefully draw attention to the various dimensions in which the smart city was rendered visible and negotiable.

Notes

1 Department of Engineering and Innovation, Open University, United Kingdom.
2 College of Design and Social Context, RMIT University, Australia.
3 Department of Engineering and Innovation, Open University, United Kingdom.
4 Department of Engineering and Innovation, Open University, United Kingdom.
5 The politics of circumstances, factors and coalitions, rather than explicit ideological notions or moral and ethical premises.

References

Albino, V., Berardi, U. and Dangelico, R.M., 2015. Smart cities: definitions, dimensions, performance, and initiatives. *Journal of Urban Technology*, 22 (1), 3–21.

Aoun, C., 2013. The smart city cornerstone: urban efficiency. Available at: www.schneider-electric.com/en/download/document/998-2095-01-30-13_EN/ [accessed 28 May 2018].

Avelino, F., 2009. Empowerment and the challenge of applying transition management to ongoing projects. *Policy Sciences*, 42 (4), 369–390.

Avelino, F., 2011. Power in transition: empowering discourses on sustainability transitions. Unpublished PhD Thesis, Dutch Research Institute for Transitions, Erasmus University Rotterdam.

Avelino, F. and Grin, J., 2017. Beyond deconstruction: a reconstructive perspective on sustainability transition governance. *Environmental Innovation and Societal Transitions*, 22, 15–25.

Avelino, F. and Wittmayer, J.M., 2016. Shifting power relations in sustainability transitions: a multi-actor perspective. *Journal of Environmental Policy & Planning*, 18 (5), 628–649.

Bakıcı, T., Almirall, E. and Wareham, J., 2013. A smart city initiative: the case of Barcelona. *Journal of the Knowledge Economy*, 4 (2), 135–148.

BIS, 2013. Smart cities background paper. Available at: www.gov.uk/government/uploads/system/uploads/attachment_data/file/246019/bis-13-1209-smart-cities-background-paper-digital.pdf [accessed 28 May 2018].

BSI, 2014. Smart city framework – guide to establishing strategies for smart cities and communities. Available at: http://shop.bsigroup.com/upload/267775/PAS%20181%20%282014%29.pdf [accessed 28 May 2018].

Brenner, N. and Schmid, C., 2015. Towards a new epistemology of the urban? *City*, 19 (2–3), 151–182.

Caird, S., Hudson, L. and Kortuem, G., 2016. A tale of evaluation and reporting in UK smart cities. Available at: www.mksmart.org/wp-content/uploads/2016/04/Tales_Smart_Cities_Final_2016.pdf [accessed 28 May 2018].

Cairns, S., Sloman, L., Newson, C., Anable, J., Kirkbride, A. and Goodwin, P., 2008. Smarter choices: assessing the potential to achieve traffic reduction using soft measures. *Transport Reviews*, 28 (5), 593–618.

Cambridge News, 2016. City Deal splashes out on new app to improve journeys in the city. Available at: www.cambridge-news.co.uk/news/cambridge-news/city-deal-splashes-out-new-12083771 [accessed 28 May 2018].

CAMK, 2014a. MK:Smart CitizenLab talks complementary transport systems in Milton Keynes. Available at: www.mksmart.org/blog/2014/05/13/mksmart-citizenlab-talks-complementary-transport-systems-in-milton-keynes/ [accessed 28 May 2018].

CAMK, 2014b. MK Smart – Redways: an asset fit for the future in Milton Keynes. Available at: http://eventful.com/miltonkeynes/events/mk-smart-redways-asset-fit-future-/E0-001-069463633-7 [accessed 28 May 2018].

Campaign for Better Transport, 2014. Car dependency scorecard 2014. Available at: www.bettertransport.org.uk/sites/default/files/pdfs/Car_Dep_Scorecard_2014_LOW_RES.pdf [accessed 28 May 2018].

Castan Broto, V. and Bulkeley, H., 2013. Maintaining climate change experiments: urban political ecology and everyday configuration of urban infrastructure. *International Journal of Urban and Regional Research*, 37, 1934–1948.

Centre for Cities, 2015. Fast-growth cities: the opportunities and challenges ahead. Available at: www.centreforcities.org/publication/fast-growth-cities-opportunities-and-challenges-for-cambridge-oxford-milton-keynes-swindon-norwich/ [accessed 28 May 2018].

Cisco, 2012. Smart city framework: a systematic process for enabling smart+connected communities. Available at: www.cisco.com/c/dam/en_us/about/ac79/docs/ps/motm/Smart-City-Framework.pdf [accessed 28 May 2018].

Cisco, 2014. Startup Spotlight: Alchera Technologies. Available at: www.cisco.com/assets/global/UK/pdfs/the_ioe_unlocking_the_opportunity_for_uk_startups.pdf [accessed 28 May 2018].

Clapson, M., 2013. *The plan for Milton Keynes*. Oxford: Routledge.

Coenen, L., Benneworth, P. and Truffer, B., 2012. Toward a spatial perspective on sustainability transitions. *Research Policy*, 41 (6), 968–979.

Connect Cambridgeshire, 2016. Smart Cambridge supports transport plans. Available at: www.connectingcambridgeshire.co.uk/smartcamb/ [accessed 28 May 2018].

Corfee-Morlot, J., Cochran, I., Hallegatte, S. and Teasdale, P.J., 2011. Multilevel risk governance and urban adaptation policy. *Climatic change*, 104 (1), 169–197.

d'Aquin, M., Adamou, A., Daga, E., Liu, S., Thomas, K. and Motta, E., 2014. Dealing with diversity in a smart-city datahub. *Proceedings of the Fifth International Conference on Semantics for Smarter Cities-Volume 1280*, 68–82.

d'Aquin, M., Davies, J. and Motta, E., 2015. Smart Cities' data: challenges and opportunities for semantic technologies. *IEEE Internet Computing*, 19 (6), 66–70.

DriveMidlands, 2016. Alchera Technologies Company Pitch. Available at: www.drivemidlands.co.uk/wp-content/uploads/2016/03/Alchera.pdf [accessed 28 May 2018].

Edwards, M., 2001. City design: what went wrong at Milton Keynes? *Journal of Urban Design*, 6 (1), 87–96.

Flyvbjerg, B., 1998. *Rationality and power: Democracy in practice*. Chicago, IL: University of Chicago Press.

Flyvbjerg, B., 2001. *Making social science matter: Why social inquiry fails and how it can succeed again*. Cambridge: Cambridge University Press.

Flyvbjerg, B., 2004. Phronetic planning research: theoretical and methodological reflections. *Planning Theory and Practice*, 5 (3), 283–306.

Foucault, M., 1979. On governmentality. *Ideology & Consciousness*, 6, 5–21.

Franklin, J., 1999. Two decades of the Redway cycle paths in Milton Keynes. *Traffic Engineering & Control*, 40 (7–8), 393–396.

Friedmann, J., 1998. Planning theory revisited. *European Planning Studies*, 6 (3), 245–253.

Gaved, M. and Peasgood, A., 2015. Location-based language learning for migrants in a smart city. *Proceedings of the 15th International Conference on Technology, Policy and Innovation*. Available at: http://oro.open.ac.uk/43797/ [accessed 15 August 2018].

Geels, F., 2014. Regime resistance against low-carbon transitions: introducing politics and power into the multi-level perspective. *Theory, Culture & Society*, 31, 21–40.

Gitelman, L. and Jackson, V., 2013. Introduction. In *'Raw data' is an oxymoron*. Cambridge, MA: MIT, pp. 1–14.

Gleeson, B., 2014. *The urban condition*. London: Routledge.

Gooch, D., Wolff, A., Kortuem, G. and Brown, R., 2015. Reimagining the role of citizens in Smart City projects. *Adjunct Proceedings of the 2015 ACM International Joint Conference on Pervasive and Ubiquitous Computing and Proceedings of the 2015 ACM International Symposium on Wearable Computers*, 1587–1594.

Greenfield, A., 2013. *Against the smart city*. New York: Do Projects.

Guy, S. and Moore, S.A., eds, 2004. *Sustainable architectures: Critical explorations of green building practice in Europe and North America*. Oxford: Routledge.

Hardy, C. and Leiba-O'Sullivan, S., 1998. The power behind empowerment: implications for research and practice. *Human Relations*, 51 (4), 451–483.

Highways Industry, 2017. Vivacity Labs Secures 3M to boost road safety for Autonomous Vehicles. Available at: www.highwaysindustry.com/vivacity-labs-secures-3m-to-boost-road-safety-for-autonomous-vehicles/ [accessed 28 May 2018].

Hollands, R.G., 2008. Will the real smart city please stand up? Intelligent, progressive or entrepreneurial? *City*, 12 (3), 303–320.

Hollands, R.G., 2015. Critical interventions into the corporate smart city. *Cambridge Journal of Regions, Economy and Society*, 8 (1), 61–77.

Hoogma, R., Kemp, R., Schot, J. and Truffer, B., 2004. Experimenting for sustainable transport: the approach of strategic niche management. *Technology Analysis & Strategic Management*, 16 (4), 561–566.

IPCC, 2014. Fifth assessment report of the Intergovernmental Panel on Climate Change – Climate Change 2014: impacts, adaptation, and vulnerability. Available at: www.ipcc.ch/report/ar5/wg2/ [accessed 30 May 2018].

Jasanoff, S., 2015. Future imperfect: science, technology, and the imaginations of modernity. In S. Jasanoff and S.-H. Kim, eds. *Dreamscapes of modernity: Sociotechnical imaginaries and the fabrication of power*. Chicago, IL: University of Chicago Press.

Kitchin, R., 2014. Making sense of smart cities: addressing present shortcomings. *Cambridge Journal of Regions, Economy and Society*, 8 (1), 131–136.

Kitchin, R., 2017. Thinking critically about and researching algorithms. *Information, Communication & Society*, 20 (1), 14–29.

Lee, J.H., Hancock, M.G. and Hu, M.C., 2014. Towards an effective framework for building smart cities: lessons from Seoul and San Francisco. *Technological Forecasting and Social Change*, 89, 80–99.

Llewelyn-Davies, W., Forestier-Walker and Bor, 1968. *Milton Keynes: Interim Report* London: MKDC.

Loorbach, D., Wittmayer, J.M., Shiroyama, H., Fujino, J. and Mizuguchi, S., 2016. *Governance of urban sustainability transitions: European and Asian experiences*. Tokyo: Springer Japan.

Miles, M.B. and Huberman, A.M., 1994. *Qualitative data analysis: An expanded sourcebook*. Thousand Oaks, CA: Sage.

Moore, T., de Haan, F., Horne, R. and Gleeson, B., 2017. *Urban sustainability transitions – Australian cases – International perspectives*. Tokyo: Springer Japan.

MKC, 2010. Milton Keynes low-carbon living strategy. Available at: www.milton-keynes.gov.uk/assets/attach/7075/Low_Carbon_Living_Strategy_-_final.pdf [accessed 28 May 2018].

MKC, 2011. Milton Keynes low-carbon action plan. Available at: www.milton-keynes.gov.uk/assets/attach/7076/Low_Carbon_action_plan_-__2011.pdf [accessed 28 May 2018].

MKC, 2012a. Employment technical paper: the Milton Keynes economy. Available at: www.milton-keynes.gov.uk/planning-policy/documents/eMPLOYMENT_tECHINCAL_pAPER_FINAL_16-05-2012.pdf [accessed 28 May 2018].

MKC, 2012b. Future-Ready MK. Available at: https://connect.innovateuk.org/documents/3130726/3794125/Feasibility+Study+-+Milton+Keynes+Council.pdf [accessed 28 May 2018].

MKC, 2013. Plugged in Places (PIP) and Electric Vehicle fact sheet. Available at: www.milton-keynes.gov.uk/assets/attach/13853/PIP_and_EV_Fact_Sheetl_Feb_2013_V2.pdf [accessed 28 May 2018].

MKF2050, 2016. Milton Keynes: making a great city greater. Available at: www.mkfutures2050.com/images/pdfs/reports/MK-Futures-2050-Full-Report-printable.pdf [accessed 28 May 2018].

MK:Smart, 2014. Spotlight interviews – John Miles. Available at: www.mksmart.org/blog/2014/10/10/spotlight-interviews-john-miles/ [accessed 28 May 2018].

Montaner, J.J.C., Kortuem, G. and Foell, S., 2015. Smart community energy initiatives in Smart Cities: learning collectively about energy. Available at: http://oro.open.ac.uk/43773/1/Learning_collectively_about_energy.pdf [accessed 28 May 2018].

Neirotti, P., De Marco, A., Cagliano, A.C., Mangano, G. and Scorrano, F., 2014. Current trends in Smart City initiatives: some stylised facts. *Cities*, 38, 25–36.

Okada, A., Wolff, A. and Mikroyannidis, A., 2015. Fostering young smart citizens through personal learning environments for urban inquiries. Available at: http://oro.open.ac.uk/44298/ [accessed 28 May 2018].

OneMK, 2016. MotionMap: free Milton Keynes app to track people and vehicles to be aunched. Available at: www.onemk.co.uk/motionmap/story-29430588-detail/story.html [accessed 28 May 2018].

Osborne Clarke, 2015. Financing the commercialisation of smart city technology. Available at: www.osborneclarke.com/media/filer_public/4e/b1/4eb1e78a-6249-4701-85d7-b9ae123c6f81/financing_the_commercialisation_of_smart_city_technology.pdf [accessed 28 May 2018].

Potter, S., Valdez, A.M., Cook, M. and Langendahl, P., 2015. Governance in niche development for a transition to a new mobility regime. *International Sustainability Conference* 2015, 25–28 August 2015, University of Sussex.

PRP Architects, 2010. Milton Keynes – a sustainable future: a low carbon prospectus. Available at: www.milton-keynes.gov.uk/environmental-health-and-trading-standards/mk-low-carbon-living/low-carbon-prospectus [accessed 28 May 2018].

Rotmans, J. and Loorbach, D., 2010. Towards a better understanding of transitions and their governance. A systemic and reflexive approach. *Transitions to sustainable development: New directions in the study of long term transformative change*. New York: Routledge.

Shapiro, J.M., 2006. Smart cities: quality of life, productivity, and the growth effects of human capital. *The Review of Economics and Statistics*, 88 (2), 324–335.

Sheffield Smart Lab, 2015. Citizen Data Portal and Trip Planner. Available at: www.sheffieldsmartlab.co.uk/finalists-city-centre-vibrancy-1/2015/11/26/citizen-data-portal-trip-planner [accessed 28 November 2017].

Shelton, T., Zook, M. and Wiig, A., 2015. The 'actually existing smart city'. *Cambridge Journal of Regions, Economy and Society*, 8 (1), 13–25.

Smith, A. and Stirling, A., 2008. Social-ecological resilience and socio-technical transitions: critical issues for sustainability governance – working papers from the STEPS Centre. Available at: www.steps-centre.org/PDFs/STEPS%20Working%20Paper_Transitions.pdf [accessed 28 May 2018].

Smith, A. and Raven, R., 2012. What is protective space? Reconsidering niches in transitions to sustainability. *Research Policy*, 41 (6), 1025–1036.

Söderström, O., Paasche, T. and Klauser, F., 2014. Smart cities as corporate storytelling. *City*, 18 (3), 307–320.

Transport Innovation Task and Finish Group, 2016. Report March 2016. Available at: http://goo.gl/gIvhZ0 [accessed 28 May 2018].

Truffer, B., Murphy, J.T. and Raven, R., 2015. The geography of sustainability transitions: contours of an emerging theme. *Environmental Innovation and Societal Transitions*, 17, 63–72.

TSB, 2013. Future cities demonstrator competition feasibility studies interim summary report. Available at: https://connect.innovateuk.org/documents/3130726/6091879/Future+Cities+Feasibility+Studies+-+Interim+Report.pdf/4ab345d6-105c-4a88-867f-12e7b976780b [accessed 28 May 2018].

TSB and ARUP, 2013. Solutions for cities: an analysis of the feasibility studies from the Future Cities Demonstrator Programme. Available at: www.arup.com/publications/research/section/solutions-for-cities [accessed 28 May 2018].

UKAuthority, 2016. BT to build on MK Data Hub for Manchester. Available at: www.ukauthority.com/smart-places/entry/6397/bt-to-build-on-mk-data-hub-for-manchester [accessed 28 May 2018].

UN, 2014. World Urbanization prospects – 2014 Revision. Available at: https://esa.un.org/unpd/wup/Publications/Files/WUP2014-Highlights.pdf [accessed 28 May 2018].

Valdez, A.M., Cook, M., Potter, S. and Langendahl, P.A., 2018. Exploring participatory visions of smart transport in Milton Keynes. *Proceedings of the Institution of Civil Engineers-Engineering Sustainability*, 171 (4).

Van Hulst, M., 2012. Storytelling, a model of and a model for planning. *Planning Theory*, 11 (3), 299–318.

Vanolo, A., 2013. Smartmentality: the smart city as disciplinary strategy. *Urban Studies*, 51 (5), 883–898.

Vilajosana, I., Llosa, J., Martinez, B., Domingo-Prieto, M., Angles, A. and Vilajosana, X., 2013. Bootstrapping smart cities through a self-sustainable model based on big data flows. *IEEE Communications Magazine*, 51 (6), 128–134.

Williamson, B., 2015. Educating the smart city: schooling smart citizens through computational urbanism. *Big Data and Society*, 2 (2).

Wolff, A., Gooch, D., Mir, U., Cavero, J. and Kortuem, G., 2015a. Removing barriers for citizen participation to urban innovation. Available at: http://oro.open.ac.uk/43854/ [accessed 30 May 2018].

Wolff, A., Kortuem, G. and Cavero, J., 2015b. Towards smart city education. In *SustainIT 2015: 4th IFIP Conference on Sustainable Internet and ICT for Sustainability*, 14–16 April 2015, Madrid, Spain. Available at: http://oro.open.ac.uk/42549/1/Sustain-OU-CR.pdf [accessed 28 May 2018].

Wolfram M., 2012. Deconstructing smart cities: an intertextual reading of concepts and practices for integrated urban and ICT development. In M. Schrenk, V.V. Popovich, P. Zeile and P. Elisei, eds. *Re-mixing the city: Towards sustainability and resilience?* Schwechat, Austria: Real Corp, pp. 171–181.

9 It's the complexity stupid!

How transition management politicises and reimagines Rotterdam's mobility system

Shivant Jhagroe and Derk Loorbach

Introduction: transition management and epistemic politics

In his 2001 book *The Moment of Complexity*, Mark Taylor states that '[a]t this point all that is solid seems to melt away, creating a sense of vertigo that is welcomed by some as the end of false consciousness and denounced by other as catastrophic nihilism' (Taylor 2001, p. 3). This impinging contingency is particularly challenging for urban governance actors. Cities have limited impact on effectively intervening in socio-technical systems, regarding for instance mobility, energy and spatial planning. One of the approaches that gained traction and has inspired policy and governance actors to grasp this complexity is 'transition management' (Loorbach 2007; Switzer *et al.* 2013). Transition management has gained increasing attention among academics and policy professionals, and impacted different actor networks in diagnosing and addressing specific sustainability challenges (Kern and Howlett 2009; Loorbach *et al.* 2017). The popularity and real-world implications of transition management for the imagination and configuration of urban systems makes it germane to critically reflect on its epistemic politics.

The underlying assumptions of transition management are grounded in complexity thinking, aiming to understand and challenge dominant socio-technical regimes and practices, while building coalitions to experiment with alternative ways of delineating, steering and operating socio-technical systems. Historically normalised system descriptions are considered inherently contingent, and therefore as subject to fundamental change. In urban environments, transition management has gained interest as well (Roorda *et al.* 2014), assuming that one or more urban socio-technical systems are 'ripe' for a transition. As transition-management interventions have been enacted in cities worldwide, it has allowed governance actors to strategically facilitate instances through which urban systems are being reconfigured (Loorbach *et al.* 2017; Jhagroe 2017).

Despite the receptivity of governance actors to this transition model, it has encountered significant criticism. Particularly interesting here is the debate between advocates and critics of transition management informed by complexity theory. Transition management critics, Shove and Walker, asked the question: who actually 'checks' the people that claim to be transition managers (2007).

These authors argue that there is a lack of attention given to the 'winners' and 'losers' of transitions, arguably because of the apolitical assumptions underlying complexity thinking. They suggest transition management expresses 'obscure politics' related to the 'high level of abstraction in its theories and practice ("systems", management of the "it")' (2007, p. 3). Rotmans and Kemp (2008), as advocates and co-developers of transition management thinking, replied to Shove and Walker with a complexity-based conception of transitions dynamics and of the role of pragmatic politics therein. They stressed the importance of pragmatic adaptation in times of reflexive modernity, while Shove and Walker referred to this as the 'politics of shape shifting' since politics is not only concerned with 'selection and variation, but also with boundary making and definition' (Shove and Walker 2008, p. 1014). This discussion illustrates that complexity theory plays a crucial role in inspiring transition-management thinking in general, as well as in very specific practices regarding system definitions and transformative agency. Despite (or rather in spite of) this conceptual flexibility and pragmatism, it seems that transition management bears the potential to reinforce planning activities, as well as advance democratic imaginations of socio-technical systems (Hendriks 2009; Audet 2012; Jhagroe and Loorbach 2015). The employment of transition-management concepts 'raise[s] questions about and prompt[s] debate in the scientific arena (for example related to normative orientation of researchers, legitimacy of interventions and lack of attention to power and politics)' (Verbong and Loorbach 2012, p. 16).

In this chapter we discuss how transition management, as a set of epistemic techniques, seeks to persuade urban-governance actors to reimagine and reshape socio-technical systems. We first discuss how transition-management concepts have been informed by complexity theory, particularly regarding notions of fundamental sustainable change and transformative agency. In cities, complexity-informed transition-management concepts can be understood as a flexible set of epistemic techniques that enable governance actors to create shared understandings of complex contexts and imagine alternative urban futures to which everyday actions can be related. An empirical case in Rotterdam (the Netherlands) is presented to explore how complexity-based transition-management concepts redefine, reimagine and – to some extent – reshape Rotterdam's mobility, including mobility policy. As such, transition management politicises Rotterdam's urban mobility systems and accommodates new system framings. Empirical material is drawn from document analysis, interviews with stakeholders, and participatory observations. Combining these methods is fruitful for engaged scholarship in the field of transition research (cf. Jhagroe 2016; Wittmayer 2016). One of the authors of this chapter was directly involved in the transition-management process and mobility-arena sessions. Active discussions with various stakeholders allowed for *in situ* understandings and knowledge co-production, which is typical for transition-management practices. The case study examines how transition-management concepts reframe and politicise mobility (policy) by calling into question car- and growth-based planning, while linking physical mobility to issues of economic opportunity and income inequality. Finally, we conclude with some reflections on the complexity of applying transition management in urban environments.

Knowing complexity and managing urban transitions

Against the background of late modernity and complex socio-economic and environmental challenges, a set of discourses and practices emerged under the banner of *sustainability transitions*. Contributors to these discourses try to develop knowledge about social transformations and conceptual tools to advance a sustainable society in the late 1990s. Based on understanding broader socio-technical shifts and societal dynamics, transition scholars started cooperating with Dutch governance actors to address current wicked problems around energy, biomass, agriculture, transport and spatial planning during the early 2000s (Rotmans *et al.* 2001; Kemp and Rotmans 2005; Loorbach 2007; Kern and Smith 2008; Rotmans and Loorbach 2010; Avelino 2011, p. 5). A broader science-policy interface, that emerged from the relationship that developed between scholars and policymakers, included 37 research programmes 'to further develop this knowledge [system innovation and transitions, SJ and DL] and put their ideas into practice with the aim of promoting transitions to a sustainable society' (KSI Network). Different transition approaches were developed in these research programmes, including 'transition management'. The development and dissemination of transition-management concepts has been pursued by academics and governance actors. Transition management is presented as a normative governance concept that enables contemporary governance actors to address and tackle the complex problems they are confronted with, framing them as systemic challenges in need of a 'sustainable transition' (Loorbach, 2010).

Complexity theory and transition management

Transition-management principles are grounded in an underlying philosophy based on complexity theory (Loorbach 2007; Rotmans and Loorbach 2009). In the last two decades, complexity theory has been introduced in a range of academic disciplines (Taylor 2001; Walby 2003; Sawyer 2005). A complexity-informed ontology presumes various 'complex adaptive systems' in nature and society, such as an ant nest, a swarm of birds, the central nervous system, a forest, a subculture, a financial system, etc. This so-called 'third wave systems theory' moves away from conceiving any system as static, highlighting their dynamic and open characteristics (Sawyer 2005, p. 10). Such complex systems are composed of small and simple elements and events, which, operating together, shape complex and dynamic patterns of adaptation, self-organisation and non-linear change. As Paul Cilliers (2002) suggests, complexity theory expresses and fits into a 'post-modern condition', in which universal knowledge claims cannot be made. Rather, epistemic grounds in a complex society are inherently partial, fluid and embedded in intricate subsystems. Even though this creates severe limits on major social and political Projects (with a capital P), understanding agency and radical change is not impossible. In fact, Connolly argues that this should make us rethink and 'appreciate multiple sites and degrees of agency' (Connolly 2011, p. 22). An ontology of complex systems, in this sense, is also a political ontology of social change. Interestingly, agency immanent in complex systems is considered fruitful for governance actors. In the context of societal (or socio-technical) transitions, Loorbach argues that:

> [t]ransition management (...) conceptualizes the role of agency in transitions and can be used to analyze possibilities for influencing. Transition management therefore necessarily builds on an understanding of transitions from a complex system perspective as basis for development of governance strategies.
>
> (Loorbach 2007, p. 18)

Complexity-informed thinking clearly plays an important role in the building-blocks of transition management. Complex systems are relatively instable and prone to interventions that allow specific attractors to develop and create tipping points towards a radically different system state. The interaction of local small actions can create emergence or emergent properties, which means changes at a higher level.

Transition management acknowledges sheer complexity and epistemic fragmentation, but does not imply steering for systemic change is impossible. Taking principles of complexity theory seriously implies knowing what complexity means and inscribing it into a governance approach for sustainable system change. According to Loorbach, transition management 'is built around the governance paradox that "societal change is too complex to handle in terms of management but it is possible to formulate relatively simple rules how to influence societal change"' (2007, p. 86). The idea of a system transition, with its predictable patterns of development, is now visualised in a so-called 'S-curve'. This curve combines ideas on innovation and diffusion with notions of path-dependency and regime destabilisation and can be used as a heuristic to explore perceived and potential dynamics of transitions. It evolved out of the S-curve that is also used in technology and market-development theories (see Figure 9.1), but was used in transition management to express underlying complexity-theoretical assumptions of long stable periods punctuated by relatively short, non-linear changes.

This pattern offers a predictive and normative scheme to intervene strategically. If potentialities of complex systems are mapped carefully, it offers a set of complexity-informed management tools (Van Raak 2016). Loorbach argues:

> The basic steering philosophy underlying transition management is that of anticipation and adaptation, starting from a macro-vision on sustainability, building upon bottom-up (micro) initiatives, meanwhile influencing the meso-regime. Goals are not fixed but developed (through a search and

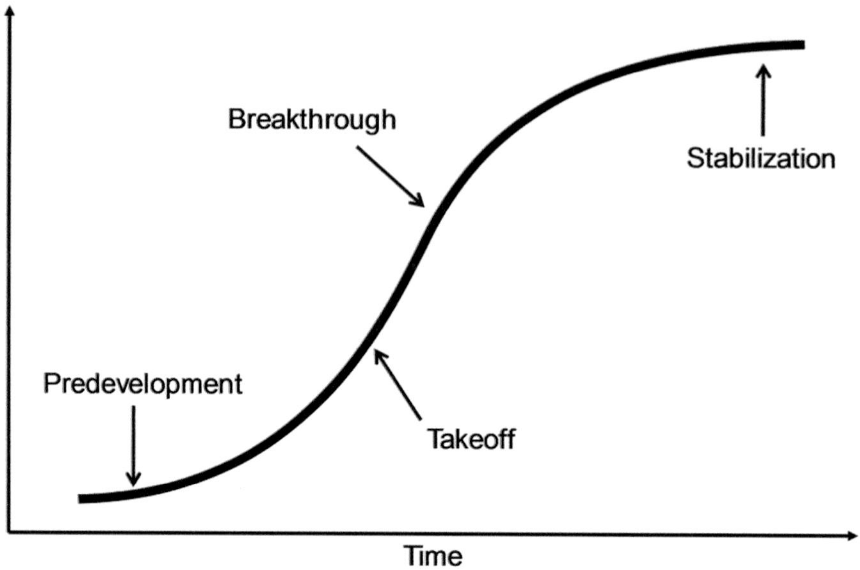

Figure 9.1 S-curve visualisation of a generic transition pattern.

> learning process) by society and the systems designed to fulfil these goals are accordingly created through a bottom up approach using incremental steps directed toward a long-term goal (e.g. directed incrementalism (Kemp 2003)).
>
> (Loorbach 2007, p. 81)

It is instructive, at this point, to briefly highlight which complexity science notions inspire transition-management concepts. Table 9.1 presents a set of complexity mechanisms, informed by a complex systems ontology, and how it has informed specific systemic instruments and management principles in transition management.

One of the most productive aspects of complexity-inspired transition management is that it can provide policymakers with insights and language to operate in fluid governance spaces when confronted with complex problems that involve multiple actors, levels and knowledge claims. This however requires a fundamentally different starting point than has been institutionalised in public administration, where the dominant perspective is policy-centric: the policy-process is the centre of solution generation and decision-making. From such an inside-out perspective, complexity is often understood as complicating processes of implementation, leading to different forms of participation or incrementalism anticipating potential surprises, resistance or complications further down the road. Transition management takes an outside-in perspective, starting from societal complexity and the possibilities it offers in terms of, for example,

Table 9.1 Inspiration from complexity science for transition management

Complexity mechanism	*Systemic instruments for TM*	*Management principles for TM*
Co-evolution	Multi-level approach, multi-domain approach	Complex systems analysis
Emergence	Creating space for niches (arenas, new coalitions)	Transition arena and pathways
Attractors	Multi-temporal approach	Sustainability visions/images
Variation	Keeping options open	Transition experiments
Selection	Learning-by-doing and doing-by-learning	Transition agenda/goals
Self-organisation	Multi-actor approach	Transition coalition and networks
Innovation of nuclei	Focus on frontrunners	Deepening, broadening and scaling up
Feedback	Reflexive governance	Monitoring and evaluation

Sources: from Loorbach 2007, p. 114; Rotmans and Loorbach 2009, p. 191.

self-organising potential, non-linearities and (social) innovation. Governance actors, then, are able to take a reflexive position in which policy is embedded in and co-evolving with a broader context. They are able to reimagine that policy box and their own role. Such a shift in perspective can have significant implications given that it also highlights the problems of the dominant institutionalised and incremental logic of policy-making. It is the basic argument behind complexity-informed transition management for a transformative role of policy in helping to support, guide and accelerate emerging new ways of thinking, doing and organising in society in a desired direction (Loorbach 2007, 2010).

Embeddedness of transition management

Furthermore, it seems that complexity-theory concepts can facilitate more inclusive governance, leading to the integration of previously unheard stakeholders and unheeded sustainability values. However, as Dean (2013) argues contemporary regimes do not oppose reflexivity, contingency and complexity, but actually feed on them. Similarly, Braun (2015) argues that there seems to be an ideologically relevant fit between dynamic ecologism, complex systems theory and flexible forms of market-oriented liberalism. These scholars argue – in different ways – that reflexivity, experimentation and complexity do not only resist capture and institutionalisation, but also express novel forms of capture and governing. This has occurred, for example, in the context of Rotterdam's sustainable waterfront transition (Jhagroe and Loorbach 2015). Here, the inclusion of mainly commercial actors reinforced the dominance of a neoliberal governing logic and the downplaying of the opinions of local residents and businesses. Clearly, the political and governance impact of transition-management concepts depends on the context in which they are deployed and the power dynamics that develop within those contexts. At the other side of the spectrum, Transition Town Networks all over the world are seeking to render self-governable their own communities – beyond regulatory and market frameworks – by experimenting with post-capitalist economic models, energy autonomy and local food systems (Jhagroe 2016).

It is also important to note that transition management is one of the approaches available to urban-governance actors to address urban complexities and contingencies. Other approaches on the 'epistemic market place' that co-exist with this particular approach are, for instance, urban resilience and actually existing incumbent sectoral knowledge configurations. More critical academic approaches such as critical urban theory and urban political ecology seek to antagonise dominant urban-system framings (cf. Jhagroe 2016). Within this landscape, the power of transition management is, arguably, to feed on existing epistemic frames and seek to challenge them 'from within', opening up spaces for sustainable system imaginaries and descriptions. In other words, 'the transition-management approach' is always subject to local experimentation, change and integration into heuristics to challenge and reframe urban systems.

The epistemic techniques of urban transition management

As is stated in the introduction chapter, transitions 'entail alternative ways of problematising these [urban] systems'. Transition management provides a set of epistemic techniques for governance, in our case, *urban* governance. Informed by insights from complexity theory, transition management seeks to render visible, knowable and governable the city (or specific urban objects) in particular ways that are so far implicit, emergent or marginalised. In this sense, transition-management knowledge and concepts resonate with what can be called an 'urban political rationality'. Resonating with the introduction chapter of this book, a political rationality (Dean, 2010), in the context of the urban, refers to a regime of governance and practices that carves out the city in a specific way and allows for particular modes of intervention. As such, a political rationality 'describe[s] particular ways of thinking about, calculating and responding to problems', and therefore 'is understood as the practical reasoning underpinning urban governance' (Chapter 1, this book). Or as Townley notes:

> Programs of government, for example, require vocabularies, ways of representing that which is to be governed, ways of ordering populations (i.e., mechanisms for the supervision and administration of individuals and groups). Rationality is dependent upon specific knowledges and techniques of rendering something knowable and, as a result, governable.
>
> (Townley 1993, p. 520)

In urban governing, modes of seeing and knowing the city – in all its complexity – are deeply entangled with particular definitions and interventions (Osborne and Rose 1999; Haahr 2004; Dean 2010). In a quite literal sense, drawings, maps, graphs, expertise, calculations, etc., can be mobilised. If high levels of carbon dioxide are considered a problem, these emissions are rendered visible and measurable (with CO_2 rates and maps), in order to evaluate trends and mitigate the perceived problem.

In the context of transition management in urban environments, the starting point is not a specific technical concern, but typically a broader urbanisation concern. It assumes a 'complex systems' way of understanding the city and its (sustainability) challenges, while mobilising complexity-inspired concepts to render knowable and govern urban spaces and subjects.

An urban transition-management guidance manual has been developed and specified for urban-governance actors and professionals (Roorda *et al.* 2014). It reports on urban transition-management projects in different European cities. Urban transition-management is a city-oriented translation of the more general theory and approach. It assumes that direct control is virtually impossible in urban transitions, yet indirect influence and experimentally exploring and developing 'sustainable urban futures' is deemed desirable and feasible. The stated goal of the guidance manual is 'to build on the innovative initiatives and inspiring visions in your city, and thus to play into and strengthen the dynamics towards a sustainable future (Roorda *et al.* 2014, p. 6). To this end, a number of actions are proposed, such as 'forming a transition team, exploring local dynamics, framing the transition challenge and envisioning a sustainable future', and more strategic follow-ups such as establishing a 'transition agenda' with strategic partners, experimentation and monitoring of progress. Note that these actions are related to the underlying 'complexity mechanism' as presented above.

A set of epistemic techniques are associated with the ways in which urban transition management is conceived and can be enacted. These epistemic techniques, or knowledge-producing practices, visualise and render knowable the city via complexity mechanisms (Dean 2010). Furthermore, they produce particular ways to see and know transformative agency for governance actors, with the overall aim to foster sustainable urban transitions. Complexity is represented in such a way that it can indeed be re-presented, and it has 'underlying mechanisms' that can be rendered knowable and used strategically. We discern four related, yet distinct epistemic techniques, which are crucial to this approach, and are associated with particular knowledge objects and underlying complexity mechanisms. Importantly, these epistemic techniques emerge in relation to a general challenge or concern, for example climate change, urban mobility or unemployment. Specific problem definitions might change, and they often do, but it provides a starting point for particular knowledge-producing techniques to work with.

Epistemic technique 1: actor analysis

A first cartography of the city is required, more particularly, describing 'relevant' local urban dynamics and frontrunners (a so-called actor analysis). Based on a mapping of the history and current developments/trends by a 'transition team', a first description can be made of the 'state of the city and its dynamics' (ibid. 2014, p. 11), in order 'to account for the complexity of the [city] we live in' (ibid., p. 20). The city is carved up in significant complex systems and interesting frontrunner vis-à-vis, for example, housing, energy or economic development. An actor analysis is particularly relevant, as it 'enables the transition team members to engage with change agents in their city, provides a structured overview of the actors related to the selected issue and to the selected objective' and filters out/in relevant actors 'for wider engagement in the later phases of the transition management approach' (ibid., p. 20). Typically, frontrunners are selected on the basis of a 'transition analysis', existing networks and snowball methods. The approach is, rather than generic actor-network analysis, specifically focused on reflexive and potentially transformative actors that represent different worldviews and system dimensions identified as relevant for a desired transition by the transition team. This technique not only leads to selection but also helps to frame participants as change agents in transition.

Epistemic technique 2: problem structuring

Before being able to proactively and strategically engage with complex dynamics in their daily contexts, problem structuring allows frontrunners to specify and establish an urban 'transition object'. Problem structuring is considered productive in settings of multiple actors, conflicting interests and uncertainties (Mingers and Rosenhead 2004). This method allows actors to frame a range of local concerns into broader systemic diagnosis, in terms of unsustainable systems that are in need of a sustainable transition. For instance, concerns of urban smog, climate change and electricity-grid peaks are framed as symptoms on underlying persistent problems inherent to the dominant system, for example a centralised fossil-energy system. This framing of problems as symptoms renders the idea of persistent problems: regular policies and approaches seek to deal with the problems but often indirectly only help strengthen the dominant system by improving it. The transition perspective based on complex-system dynamics then suggest that such persistent problems ultimately will lead to non-linear systemic change as both the contexts changes, incumbent regimes cannot adapt and change agents will develop better alternatives. Diagnosing the city and articulating a proper object of transition, is enacted by frontrunners in so-called 'transition arenas'. Transition arenas are spaces in which convergences takes place 'of the various problem perceptions [which are] facilitated from the articulation of diverging perspectives of the actors involved, which in turn will lead to new insights into the nature of the problem(s) and the underlying causal mechanisms' (Loorbach 2010, p. 174). Problem structuring offers a clear epistemic link to the urban 'transition challenge' deemed relevant.

Epistemic technique 3: envisioning

Following the problem-structuring method, which establishes urban concerns as transition concerns, sustainable urban futures are envisioned. That is to say, the unsustainable city or specific unsustainable urban systems are contrasted with ideal(ised) sustainable urban systems. This is done by exploring and visualising different desirable (social and greener) urban futures, again, in transition arena spaces. These spaces are also 'an innovative participatory process of envisioning, searching, learning and agenda-building aimed at social learning as a means to achieve (sustainable) social change' (Loorbach 2007, p. 44). Importantly, as some have pointed out, exactly the open and idealistic character of long-term sustainability visions allows for consensus and legitimacy (Kenis *et al.* 2016). This imaginary and epistemic convergence establishes 'a common language and sense of direction' (Roorda *et al.* 2014, p. 27) and is expressed by discussing possible sustainable urban futures, and elaborating them in more detail. This often takes the form of particular storylines, drawings and imagining everyday life and work in the idealised sustainable city. If envisioning is not preceded by the problem-structuring phase, it can often lead to inspiring yet non-transformative visions. Visioning in transition management is rather about identifying desirable system conditions that might help guide short-term actions to navigate through highly uncertain, unpredictable and chaotic changes.

Epistemic technique 4: backcasting

A fourth epistemic technique is to render knowable transformative agency for governance actors. Agency is already suggested in the transition arena (see above), but also in complex urban systems that are dynamic and mouldable. To this end, long-term goals about sustainable energy systems or a climate neutral city, are connected to short-term actions and translated into more manageable transformative actions. Particular storylines and 'transition paths' are developed 'describing a possible route from the present towards the envisioned future' (ibid., p. 30). This 'back-casting' refers to formulating a 'transition agenda' with local partnerships and actors (mid-term), but also doing experimental projects and evaluating them (short-term). The combination of these governance activities is discussed and planned, so actors know what to expect and to do. These modes of transformative agency assume far-from-equilibrium urban systems that are sensitive to strategic interventions towards a system 'tipping point'. The function of backcasting is both to identify potential turning points and intermediate goals, as that has a psychological and strategic effect. It allows participants to accept and accommodate a diversity of pathways simultaneously working in the same direction. It helps to internalise the notion of variation and for creative competition, as well as the philosophy of 'learning-by-doing and doing-by-learning'.

What these epistemic techniques do is to utilise complexity thinking in such a way that it renders knowable, in very particular ways, the city and its dynamics. They fixate specific knowledge objects (frontrunners, unsustainable

Table 9.2 Epistemic techniques and complexity concepts of urban transition management

	Epistemic technique	*Epistemic object*	*Inspirational complexity mechanisms*
1	Actor analysis	Change agents	Complex adaptive systems, co-evolution, emergence, self-organisation
2	Problem structuring	Persistency and symptoms of unsustainability	Patterns and dynamics of transitions
3	Envisioning	Guiding sustainability principles, sustainable urban future(s)	Attractors, selection
4	Backcasting	Short/mid-term actions, tipping points, experimentation	Variation, co-evolution, self-organisation, feedback

systems, sustainable futures, and actions), which would then make urban transitions governable. It fixates specific knowledge objects (frontrunners, unsustainable systems, sustainable futures, and actions), which are said to make urban transitions governable. An urban transition management framework is considered strategic in the sense that it does not assume a blank slate. Governance actors can connect an urban system analysis to 'an existing CO2 baseline study, and the transition agenda can inspire policy formulation' (Roorda *et al.* 2014, p. 11). This allows for pragmatic and deliberative ways of visualising and knowing opportunities for an urban transition, inspired by complexity thinking. With these forms of knowing the urban, governance actors are able to foreground a different and very specific 'diagrammatic' of the urban fabric (Osborne and Rose 1999, p. 737). It allows them to understand socio-technical systems and changes differently, and, consequently, govern them differently.

The knowledge-producing repertoire of urban transition-management is not considered a step-by-step sequential plan. If city authorities already run a pilot project, for instance, they can embed it in a wider approach of strategic experiments. Or, if an urban sustainability agenda has already been formulated, it might be complemented with various transition pathways and transition experiments (Roorda *et al.* 2014, p. 15). This modular approach underscores the openness of transition-management concepts, and their flexible applicability in governance settings (see Table 9.2). Having presented the epistemic techniques associated with urban transition management, we now turn to the empirical case.

Empirical case: politicising and reimagining Rotterdam's mobility

This section analyses a transition-management process in the domain of Rotterdam's mobility system from a knowledge-politics and epistemic framing perspective. In a transition arena process (2015–2017) the above-described approach was followed alongside the regular policy process to develop a new urban-mobility plan.

For decades, the city of Rotterdam's mobility system has been car-centred. Rotterdam is marked by its grim history, in that its city centre was bombed during the Second World War and had been rebuilt according to a modernistic vision: broad streets and modern architecture. This has shaped Rotterdam's epistemic regime as well. The layout had clearly facilitated car use and the city is heavily dominated by automobility. Infrastructurally, as well as in terms of car culture, the city of Rotterdam had been developed towards accessibility for vehicular traffic and economic development. A well-developed infrastructural network was established and an institutional system was developed around automobility and public transportation. For decades, public and political debates were dominated by the sentiment that cities indeed should be accessible to all and mode of transport should be an individual choice, rather than something to be steered. For decades, the epistemic regime that has informed Rotterdam's modernistic transport planning was based on growth, optimisation, safety, improved flow for motorised traffic and the use of traffic models. Clearly, this has not been a purely 'modern Dutch experience', as governing urban mobility by privileging motorised transportation has been a European trend (cf. Jeekel 2016).

Around 2010, the situation started to change. On the one hand, health issues of fossil-based mobility and congestion issues resulted in questions being raised about the hegemony of the car. Furthermore, a significant increase in cyclists started to become visible while car use was stagnating and a number of newly built parking garages remained half empty. The parking places were built according to dominant 'parking norms' (e.g. '2.1 parking space units' per household), but apparently inner-city inhabitants were less likely to buy cars. Combined with reflecting upon larger issues such as climate change and fossil dependence, this led to discussions in the transition team about whether these observations could be interpreted as early signals of 'a transition'. Further discussing how policy itself related to the historically developed automobility regime, and responded to these external changes, the civil servants involved started to acknowledge that their own urban policies had, in fact, helped to make cars dominant. While, formally speaking, each mode of transport has been accommodated 'equally', the fact that cars caused most direct problems (e.g. pollution, safety, congestion) that needed solving, meant they were in practice dealing with car-related issues most of their time.

Actor analysis of Rotterdam's mobility system

In the beginning of 2015, a group of civil servants approached DRIFT at the Erasmus University to discuss the possibilities of organising a transition-arena process for mobility. One of the civil servants involved had previously worked at a district level and experienced a transition-arena process there. DRIFT and

Rotterdam's municipality have cooperated often. This time, they were about to start an administrative process to develop the new five-year mobility programme for the city of Rotterdam. After a few discussions, a plan was made to complement the mobility programme (with sub-programmes on issues such as parking, congestion, biking, infrastructure and public transport) with a trajectory of a number of 'mobility arenas'. It was discussed with and supported by the alderman responsible for mobility, who referred specifically to one of the outcomes of a previous transition arena DRIFT had co-organised in Ghent, Belgium, called the *Leefstraat* (Living Street).

The transition-management trajectory started with establishing a 'transition team', consisting of three civil servants, a student studying for a master's degree and three DRIFT action researchers. Together they started to rethink and politicise Rotterdam's mobility history, its current regime, and identify potential mobility-centric transitions and transition actors. Some key arguments for change were mapped, including the rise of urban health discourse, criticism against car dominance, as well as economic inequality. Simultaneously, a number of significant trends related to more sustainable modes of transportation and the rise of the 'sharing economy' were mapped. The transition team then interviewed some 30 individuals with very diverse backgrounds that were entrepreneurial and engaged with alternative mobility futures in the broadest sense. They were identified during the analysis by snowballing or through the networks of the transition team members. These change agents were selected on the basis of their knowledge and/or innovative ideas regarding a fundamental change of Rotterdam's mobility system. From these interviews 15 were invited to join the transition-arena sessions, selected based on diversity of background, perspectives and potential roles.

Problem structuring of Rotterdam's mobility system

The first arena meetings started by presenting the transition analysis of the transition team. It framed the situation of mobility in Rotterdam at that time in transition terms: a hegemonic automobility regime that emerged historically but now was in a lock-in and causing many unsustainability effects. Policy was only addressing these 'symptoms', but alternative niches were emerging. The core-problem definition that emerged during these two arena sessions and the facilitated discussion in the group built upon foregrounding the participants' experiences, ideas and perspectives. It developed around an understanding of a dominant car-based mobility system as excluding alternatives, reinforcing individualism and societal disconnect, and as being inherently unsustainable in the long run. Combined with the potential and rapid development of alternative technologies, practices, ambitions and visions (ranging from electric mobility alternatives, to sustainable-area development to new logistics to biking), this helped the actors to challenge the dominant optimisation logic and call into question Rotterdam's mobility system as a whole. It opened up the discussions from thinking about mobility in purely physical, technological, environmental and functional terms to include socio-economic aspects and forms of inequality.

This new narrative coalesced during the second and third meeting (see Figure 9.2) around the notion of mobility poverty, referring to the lack of access to diverse modes of transport because of low income or cultural norms. It also referred to the lack of destinations or places to go because of unemployment or social exclusion. In the first category, participants cited examples and experiences that highlighted the lack of cycling culture among many minority/ethnic groups, and a negative attitude towards cycling amongst particular youth groups. Many people were either limited to public transport or cars, which are relatively expensive for lower-income groups, or to walking. The effect was that many people were only physically experiencing their own neighbourhoods rather than different parts of the city. One arena participant shared the story of how he took Moroccan and Turkish teenagers on their first bike-rides for a so-called urban safari to the richer parts of the city. It was the first time these teenagers, who were born in Rotterdam, had been to this part of the city.

Policymakers reflected upon the inherent tendency in mobility policy to focus on improving Rotterdam's existing mobility regime through modelling, infrastructural provisions and control and safety measures and risk reduction strategies. This type of mobility policy has a very strong tradition of long-term planning, but in everyday practice mainly focuses on reactive responses to emerging problems; for example, by designating more space for cars to deal with congestion and growth in car use, creating separate car lanes to improve safety, or giving cars more priority to improve the traffic flow. Such actions actually

Figure 9.2 Discussing Rotterdam's mobility system in transition arena.

reinforce car-dominance and limit the space for alternatives, thereby through everyday practice reinforcing the existing automobility regime. This led policy actors to openly question the dominant policy discourse that each modality was being treated equally and to start to think about a different operational logic.

Envisioning sustainable and inclusive mobility in Rotterdam

After discussing and politicising car dominance and mobility in Rotterdam, a new narrative emerged over the course of five arena sessions. Actively reconceiving mobility challenges and system descriptions took place in a public-society platform that led to a vision of mobility that targeted mobility poverty and promoted an inclusive and sustainable city. In particular, the arena sessions helped to challenge the focus in mobility planning from solving the problems that cars create, and exploring ideas that solve problems cycling and public transport face. This also included using cycling as a means to emancipate marginalised groups in non-privileged areas of Rotterdam.

During vision- and strategy-oriented discussions in the transition arena, and using the S-curve as a heuristic (see Figure 9.3), policy actors explored and defined a new policy logic in terms of giving priority to solving problems of sustainable mobility like biking, walking, public transport, or electric and shared mobility. The themes and topics discussed were based upon existing mobility-policy

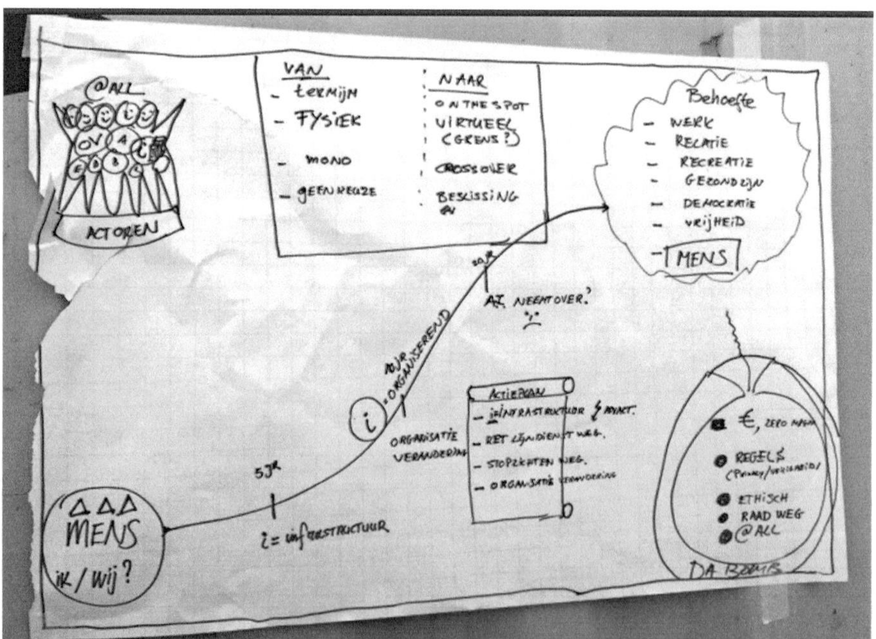

Figure 9.3 Envisioning a more sustainable and inclusive mobility system.

challenges like congestion, parking, pollution and accessibility. The transition vision articulated in the transition arena facilitated the formulation of ideas and strategies to actually create problems for policy to solve by mobilising people to bike more, to introduce shared electric mobility, to develop more affordable and high-density water-mobility systems, and transform public space by developing greener, denser and more walkable neighbourhoods.

The transition-arena process produced a new vision for mobility, which was documented in a publication titled 'New Ways' in 2015 (in Dutch: *Nieuwe Wegen*) (Municipality of Rotterdam, 2016). This document contained the epistemic framings developed during the transition arena process, including the ideas for more specific actions and interventions. It was presented to the alderman responsible for urban mobility and interested public. Two specific elements combined to provide a new perspective for the mobility systems and mobility policy in Rotterdam: (1) how the city itself should be *a place to be and to go* and where mobility is facilitating this in terms of physically helping people and goods to move around, i.e. mobility as a means not a goal; and (2) creating and *connecting places* where people are and want to be (Municipality of Rotterdam, 2015).

Backcasting for sustainable and inclusive mobility in Rotterdam

Envisioning a more sustainable and inclusive mobility system was translated into specific transition pathways described under such titles as clean mobility, sharing, mobility happiness, living public spaces and new mobility services. More actors were engaged in exploring what these transition scenarios might involve and a number of transition experiments were formulated for cycling, electric mobility and car-sharing, clean transport over water, living streets. There were also initiatives to develop new socio-economic conditions through which people could engage in different ways with mobility. During the mobility arena, and through the engagement of other civil servants from the mobility programme and urban society at large, a lot of the ideas and elements from the transition agenda were adopted within the official mobility programme of the city. On the basis of this agenda, a number of short-term experiments were adopted by the mobility department, including cycling initiatives for residents from the southern part of the city and enhancing access to art amongst low-income groups. Many of the experiments were taken further by various participants. Additionally, the knowledge-producing technique of backcasting established a number of more specific and tangible scenarios and projects that informed short-term action. The complexity associated with the broader vision of sustainable and inclusive mobility in Rotterdam (moving away from a car-centric policy) was translated into more tangible governance practices that applied a range of ideas experimentally.

Impact and limits of transition-management's epistemic techniques

The case study highlights how complexity-informed transition management enabled urban governance actors to challenge and redefine, on paper at least, Rotterdam's mobility system. But, how did the epistemic techniques associated with transition management impact upon actors and governance? And, how extensive was the impact of knowledge-producing techniques?

Impact became visible in the mobility case almost two years after the mobility-arena sessions took place. A new narrative about inclusive and sustainable mobility had spread widely, seemingly co-evolving with a sharp rise in cyclists and the use of public transport, while car use remained stable or perhaps declined very slightly. The narrative stimulated ideas for phasing out car parking in public spaces, pushing automobility outside the city centre and introducing emission zones. It has also directly reshaped the nature of policy-making in the mobility domain, in that a more experimental and pragmatic way of working came about. Policy initiatives were set up, such as widening bike-lanes, experimenting with heat and rain sensors at bike-traffic lights to determine traffic-light priorities, offering inhabitants of streets the possibility to close off their street for two months to use it as living street, and a social cycling programme seeking to develop local professions, teaching people to bike to offer freedom of mobility. The policy department started to experiment with new parking rules and offered project developers the possibility to lower parking norms for new project development voluntarily if car-sharing and sustainable mobility plans were implemented.

Participants in the mobility-transition arena were empowered to further develop initiatives they were already working on and/or develop new experiments, sometimes in collaboration with civil servants, but also based on their own initiative or entrepreneurial expertise. These initiatives included, for instance, a project to empower 500 women with multi-ethnic backgrounds to learn how to bike, which was one year later expanded with a 'bike-bank', collecting bikes and accepting bike-donations to give out for free to lower-income groups. Projects developed to temporarily use parking spaces for gardening or as recreational space, and gradually broadened into experiments under the header of 'Happy Streets', also involving designers to intervene in the public space, using paint to demarcate (temporary) bike lanes. The water taxi managed to develop a public-transport licence and offer lower rates and recently expanded with a social bike-sharing system connected to the water taxi, called 'ROS' (Rotterdamse Stadsfiets). Furthermore, a number of larger policy programmes have been launched, such as *Fietsen op Zuid* (Cycling in Rotterdam South) in which societal actors work together with transport and social policy departments to stimulate cycling, educate people to cycle, develop jobs in cycling logistics and organise workshops and cycling lessons at all the schools.

The new narrative and network obviously did not directly or completely change the dominant policy-centric and optimisation-focused practice. Institutional incrementalism appears difficult to bypass other than through enthusiastic entrepreneurial and bottom-up activity. A good example is the Rotterdam-based bike-sharing scheme that was initiated by a few participants in the transition arena and further developed beyond the project. During the last two years, these participants have been negotiating with the municipality for bike stations to be installed at more than 300 locations across the city. While still discussing this, the city was confronted with a large number of bike-sharing schemes that do not have specific stations but are 'footloose' (five by the end of 2017). This practically meant bikes appeared all over the city, occupying and sometimes littering public space to serve profit-seeking international companies. But this in itself helped the case for a more social, local and managed system, so that by mid-2018 ROS was founded. So, a new frame of urban mobility governance has arguably emerged, which focuses on sustainable and inclusive public mobility, and includes both physical and socio-economic mobility considerations. If anything, Rotterdam's mobility system appears to have opened up, in part to render the system more sustainable and healthy, but also to become more inclusive and to emancipate marginalised groups.

Concluding remarks

This chapter examined how epistemic techniques associated with transition management challenged and, arguably, transformed systemic frames and urban governance in the Rotterdam case study. We showed how various epistemic techniques embedded in transition management, and linked to complexity theory, helped actors to construct new urban realities that, at least to some extent, have reshaped how Rotterdam's mobility systems is understood. We illustrated how new knowledge frames emerged that contrasted with the previously dominant political urban-management and planning rationalities. Alternative forms of 'knowing Rotterdam' emerged in parallel with a new way of problematising mobility planning. As such, epistemic interventions through transition management constitute what Michel Foucault (1980) called counter-knowledge. Our analysis indicates that transition-management interventions have transformed some of the language and knowledge used in Rotterdam's prevailing institutionalised governance practices. The new narratives and sustainability visions were developed in such a way that they were internalised, enriched and started to diffuse through the daily practices and interactions of the participants. Thereby they led to a discursive change that is increasingly enacted through practices and made tangible through new artefacts and physical changes in the urban context. An epistemological shift can be understood as the introduction of a new way of imagining and thinking about urban systems and policy. This is symbolised by shifts, in our case study, from a car-centric to an increasingly bike-centric and socio-culturally aware political rationality. None of this would work if the narrative or facilitation was not accepted. Yet, at the same time the process was deliberately steered towards certain outcomes: the need for an urban sustainability transition. Including typically marginalised or underrepresented governance actors, and reformulating systemic challenges, long-term ambitions and short-term actions, can indeed serve as a means to encourage actors to question hegemonic epistemic frames for urban governance. Especially the complexity-informed model of strategic experimentation turned out to alter how governance actors understood, and steered change of, socio-technical systems. Any failure or limited impact, therefore, is part of the deal, for they articulated a form of governance by experimentation and local adjustment.

However, epistemic transition interventions that advocate urban sustainability do not automatically have a moral high-ground, simply because they support regime change by representing social complexity and epistemic fragmentation. As we have seen, the ways in which urban systems are observed, defined and become governable are indeed subject to struggle and negotiations. Deploying complexity theory in urban governance remains a struggle over knowledge claims. The interesting hypothesis we can formulate based upon the empirical examples presented in this chapter is that with the right timing and mobilisation of counter-hegemonic knowledge and power, it is possible to break through regime barriers. A transition-management approach, and corresponding epistemic interventions, when employed in a context where persistent problems occur and a growing number of actors start to explore alternative futures, can accelerate and guide such an emerging transition by exploring and introducing new epistemic frames; such a frame is a synthesis of elements already present, that, by articulating them, empowers those that subscribe to it and makes visible to others what the potential of an urban transition is. In doing so, more governance actors might be able to engage with the new frame and act upon it, thereby gradually making it the new reality.

References

Audet, R., 2012. Transition as discourse and the issues of interventionism, justice and innovation. In *ISEE2012 Conference – Ecological Economics and Rio*, Vol. 20, pp. 16–20.

Avelino, F., 2011. *Power in transition: Empowering discourses on sustainability transitions*. Doctoral thesis, Dutch Research Institute for Transitions (DRIFT), Erasmus University Rotterdam.

Braun, B., 2015. The 2013 Antipode RGS-IBG Lecture: new materialisms and neoliberal natures. *Antipode*, 47 (1), 1–14.

Cilliers, P., 2002. *Complexity and postmodernism: Understanding complex systems*. London: Routledge.

Connolly, W.E., 2011. *A world of becoming*. Durham, NC: Duke University Press.

Dean, J., 2013. Complexity as capture: Neoliberalism and the loop of drive. *New Formations: A Journal of Culture/Theory/Politics*, 80, 138–154.

Dean, M., 2010. *Governmentality: Power and rule in modern society*. London: Sage.

Foucault, M., 1980. *Language, counter-memory, practice: Selected essays and interviews*. New York: Cornell University Press.

Haahr, J.H., 2004. Open co-ordination as advanced liberal government. *Journal of European Public Policy*, 11 (2), 209–230.

Hendriks, C.M., 2009. Policy design without democracy? Making democratic sense of transition management. *Policy Sciences*, 42 (4), 341–368.

Jeekel, H., 2016. *The car-dependent society: A European perspective*. London: Routledge.

Jhagroe, S., 2016. Urban transition politics: how struggles for sustainability are (re)shaping urban spaces. DRIFT, Erasmus University Rotterdam, Rotterdam.

Jhagroe, S. and Loorbach, D., 2015. See no evil, hear no evil: the democratic potential of transition management. *Environmental Innovation and Societal Transitions*, 15, 65–83.

Kemp, R. and Rotmans, J., 2005. The management of the co-evolution of technical, environmental and social systems. In M. Weber and J. Hemmelskamp, eds. *Towards environmental innovation systems*. Berlin: Springer, pp. 33–55.

Kemp, R. and Rotmans, J., 2009. Transitioning policy co-production of a new strategic framework for energy innovation policy in the Netherlands. *Policy Sciences*, 42 (4), 303–322.

Kenis, A., Bono, F. and Mathijs, E., 2016. Unravelling the (post-)political in transition management: interrogating pathways towards sustainable change. *Journal of Environmental Policy & Planning*, 18 (5), 568–584.

Kern, F. and Howlett, M., 2009. Implementing transition management as policy reforms: a case study of the Dutch energy sector. *Policy Sciences*, 42 (4), 391.

Kern, F. and Smith, A., 2008. Restructuring energy systems for sustainability? Energy transition policy in the Netherlands. *Energy Policy*, 36 (11), 4093–4103.

KSI Network [www.ksinetwork.nl/what-is-ksi/about-ksi] (visited in 2017 and 2018).

Loorbach, D., 2007. *Transition management: New mode of governance for sustainable development*. Doctoral thesis, Dutch Research Institute for Transitions (DRIFT), Erasmus University Rotterdam.

Loorbach, D., 2010. Transition management for sustainable development: a prescriptive, complexity-based governance framework. *Governance*, 23 (1), 161–183.

Loorbach, D., Frantzeskaki, N. and Avelino, F., 2017. Sustainability transitions research: transforming science and practice for societal change. *Annual Review of Environment and Resources*, 42.

Loorbach, D. and Rotmans, J., 2010. The practice of transition management: examples and lessons from four distinct cases. *Futures*, 42 (3), 237–246.

Mingers, J. and Rosenhead, J., 2004. Problem structuring methods in action. *European Journal of Operational Research*, 152 (3), 530–554.

Municipality of Rotterdam, 2016. *Nieuwe wegen inslaan. Mobiliteit als katalysator voor een duurzame toekomst van Rotterdam*. Rotterdam: Municipality of Rotterdam.

Osborne, T. and Rose, N., 1999. Governing cities: notes on the spatialisation of virtue. *Environment and Planning D: Society and Space*, 17 (6), 737–760.

Roorda, C., Wittmayer, J., Henneman, P., van Steenbergen, F., Frantzeskaki, N. and Loorbach, D., 2014. Transition management in the urban context. Guidance manual. DRIFT, Rotterdam, Erasmus University Rotterdam.

Rotmans, J., Kemp, R. and Van Asselt M., 2001. More evolution than revolution: transition management in public policy. *Foresight*, 3 (1), 15–31.

Rotmans, J. and Kemp, R., 2008. Detour ahead: a response to Shove and Walker about the perilous road of transition management. *Environment and Planning* A, 40 (4), 1006–1012.

Rotmans, J. and Loorbach, D., 2009. Complexity and transition management. *Journal of Industrial Ecology*, 13 (2), 184–196.

Rotmans, J. and Loorbach, D., 2010. Towards a better understanding of transitions and their governance: a systemic and reflexive approach. In *Transitions to sustainable development: New directions in the study of long term transformative change*. New York: Routledge, pp. 105–198.

Sawyer, R.K., 2005. *Social emergence: Societies as complex systems* Cambridge: Cambridge University Press.

Shove, E. and Walker, G., 2007. CAUTION! Transitions ahead: politics, practice, and sustainable transition management. *Environment and Planning* A, 39 (4), 763–770.

Shove, E. and Walker, G., 2008. Transition management and the politics of shape shifting. *Environment and Planning* A, 40 (4), 1012–1014.

Switzer, A., Bertolini, L. and Grin, J., 2013. Transitions of mobility systems in urban regions: a heuristic framework. *Journal of Environmental Policy & Planning*, 15 (2), 141–160.

Taylor, M.C., 2001. *The moment of complexity: Emerging network culture*. Chicago, IL: University of Chicago Press.

Townley, B., 1993. Foucault, power/knowledge, and its relevance for human resource management. *Academy of Management Review*, 18 (3), 518–545.

Van Raak, R., 2016. *Transition policies: Connecting system dynamic, governance and instruments in an application to Dutch healthcare*. Doctoral thesis, Dutch Research Institute for Transitions (DRIFT), Erasmus University Rotterdam.

Verbong, G. and Loorbach, D., eds, 2012. *Governing the energy transition: Reality, illusion or necessity?* New York: Routledge.

Walby, S., 2003. Complexity theory, globalisation and diversity. Conference of the British Sociological Association, University of York, York.

Wittmayer, J.M., 2016. *Transition management, action research and actor roles: Understanding local sustainability transitions*. DRIFT, Rotterdam.

10 Conclusions and perspectives

Jens Stissing Jensen,[1] Matthew Cashmore[2] and Philipp Späth[3]

On the knowledge politics of urban-transition governance

Interest in the politics of urban sustainability transitions has burgeoned in recent years (e.g. Partzsch 2017; Patterson *et al.* 2017; Temper *et al.* 2018) and growing attention is being given to the role of political power in conceptualising and governing radical urban change (see Chapter 2, this volume).The purpose of this book has been to extend such discussions by drawing attention to one particular feature of the politics of transitions, knowledge politics.

The contributors to this book share a conception of knowledge as a fundamental, but irretrievably political, component of urban sustainability governance. This perspective rejects as naive a belief that knowing contributes to governance by accurately mirroring urban reality in the guise of words, drawings, calculations and numbers. Rather, we have argued that all representations of urban realities involve purifying, delineating and arranging real-world complexities into defined concerns, objects and systems. Knowing thereby contributes to structuring collective interpretations of reality in such a way that the urban fabric can be apprehended as predictable and actionable. A knowledge-politics perspective thus implies that interest-driven struggles over 'who gets what, when and how' (after Laswell 1935, see Chapter 2) are rooted in a more primordial politics, centred around how actors, concerns, technologies and resources are made visible as knowable components of urban reality.

The knowledge-politics perspective explored in this book is consequently based on an assumption that actors are inevitably influenced by the concepts, vocabularies, numbers and narratives that they have at their disposal to render their reality intelligible. Knowledge politics refers, in part, to how urban reality is constructed, controlled and sometimes re-crafted to produce particular conceptions of our collective reality and for shifting purposes.

From this point of departure, our goal with this book is to draw critical attention to and explore the particularities of knowledge politics. In so doing, the contributors to this book have mapped governance landscapes comprised of: multifarious knowledge-configuring devices and processes; situated visibilities, and their corollary, invisibilities; political rationalities; empowerments and disempowerments; and, contestation over knowledge in terms of ownership, access and its perceived relevance to governance.

In Chapter 2, for example, we explored how recourse to different forms of knowledge (knowledge rendered as art and knowledge rendered as economic calculations) was central to enrolling the support of particular audiences in the redevelopment of the derelict High Line elevated railway in New York City. Flyvbjerg's (1998) argument that power is knowledge, for '[o]ne of the privileges of power … is the freedom to define reality' (p. 229), has also been explored in various ways in this book. The example of the High Line Park illustrates how the societal elite (albeit one sometimes portrayed as being part of a 'grassroots' movement) defined and subsequently redefined the ways in which this redevelopment was conceivable, with important consequences for the eventual design, the restrictions placed on what uses of the park were allowed, and who benefited from the development.

In the case of Smart City Barcelona (Chapter 3, this volume), we saw how a political quest to create an international profile in smart-city technology could render invisible the needs of local communities, with a community food bank being displaced by a smart-city administrative hub in a city beset by mortgage foreclosures and associated economic and social problems. Nevertheless, what this case also shows is that political rationalities are evidently not insuperable. Contestation over whose reality counts (Chambers 1997) did not lead to the reproduction of the status quo in this instance; indeed, the case illustrates that power relations may be more dynamic than much of the writing on power would seem to indicate. Nor are expert cultures, and the power relations encompassed therein, fixed across space and time. As Jasanoff (2005) shows, while scientific knowledge remains a hegemonic force in Europe and North America, the precise ways in which scientific knowledge and its experts influence discourses varies between the political cultures of nation states. The contributors to this book have shown that knowledge politics are especially rich and contested in urban environments. It is within this dynamic milieu, we suggest, that an important lever for navigating sustainability transitions may lie.

The 'nuts and bolts' of urban knowledge production

The knowledge politics of urban transition governance may take many forms. Extant literature on transition governance has highlighted the construction of radical longsighted visions of systemic change as a particularly relevant epistemic technique in the governance of urban change (see e.g. Chapter 9, this volume). In this book we have argued for including a much broader range of epistemic techniques to fully appreciate the knowledge politics of urban transition governance. More specifically, we have drawn attention to those knowledge devices implicated in everyday urban governance and planning, because the political rationalities that inform day-to-day planning are important in ensuring systematic and widespread transformations of urban reality. We consequently propose that the mundane devices and procedures that, often subtly, configure our everyday knowledge about the urban fabric constitute a key site of urban transition politics.

One observation from the previous chapters is that some knowledge devices employed to shape urban governance are more likely than others to promote transition-oriented political rationalities. In Chapter 5, for example, it was demonstrated that knowledge devices such as 'best-practices' tend to systematically obscure comprehensive, systemic perspectives on the city by rendering the urban fabric visible as individualised practices and technologies. One important dimension of urban knowledge politics is thus related to whether urban reality is made conceivable in the guise of individualised components and solutions, or whether knowledge production is configured to recognise systemic reciprocities between practices, technologies, infrastructures and regulations as relevant objects of governance.

Another political aspect of knowing relates to the construction of system boundaries. Take as an example the influence of mass-flow analysis on the governance of the waste system in Mexico City (Chapter 7, this volume). This case illustrates how only those waste flows that are currently under the direct control of the government are included in the mass-flow analysis. This delineation of 'relevant' waste flows is instrumental in rendering invisible the many informal waste-handling practices of the system and the related sustainability concerns. By cementing the divide between the governed and un-governed parts of the waste-handling system in Mexico City, mass-flow analysis furthermore contributes to enacting a shortsighted political rationality, as this way of knowing closes down opportunities for considering alternative ways of defining the long-sighted concerns pertaining to the governance of the system. A fundamental political aspect of knowing is accordingly the configuration of particular temporal orientations for urban governance, and attendant questions of who benefits and who loses from particular temporal orientations. A further example of how knowledge production contributes to steering temporal orientations is the use of longsighted socio-economic price signals in the appraisal of heat-infrastructure investments in Denmark (see Chapter 6, this volume). The case illustrates how the price signals are constructed on the basis of particular assumptions that favour the status quo and high-carbon-energy-system scenarios. The calculation of price signals thus functions as a knowledge device that prevents heating utilities from investigating politically attractive low-carbon-energy-system futures, as the price signals for such futures are rendered opaque.

Yet another critical aspect of contemporary knowledge politics concerns the impact of sustainability as a guiding concern of urban system governance. The contributors to this book in different ways illustrate how sustainability is made knowable as a relevant concern: for example, in the guise of the calculated socio-economic benefits of the health effects related to cycling, as physical mass flows of waste and as transport inequalities related to car-centric mobility systems. The contributors also illustrate that sustainability seldom constitutes an integrated, overarching and coordinating way of knowing the city. Rather, sustainability tends to appear as a loosely organised network of social and bio-physical concerns, that are selectively mobilised to provoke and modulate established ways of knowing. We thus see little evidence of the emergence of an overarching new sustainability-oriented knowledge paradigm, but instead particular sustainability concerns are mobilised to modulate preconfigured landscapes of political rationalities embedded in established knowledge practices and knowledge devices.

Re-coding urban knowledge production

Given that the devices and procedures involved in the routine machinations of urban knowledge production control what aspects of urban reality are rendered visible as germane objects of governance, the re-coding of these devices and procedures is an important aspect of transformative urban governance.

The chapters in this book illustrate how the configuration of urban knowledge production may sometimes be influenced by overt interests. Take as an example the various controversies over how to render the transport system visible using smart-city technologies in the car-dominated British city of Milton Keynes (Chapter 8, this volume). This case shows how plans to render 'business' visible as a characteristic of the transport system were opposed by retailers, based on fears that this visibility might allow citizens to reconfigure their daily commuting and shopping patterns in ways that the retailers were uncomfortable with. The identification of these tensions, and the associated power dynamics, illustrates how new ways of knowing may conflict with the interests of influential actors. This case also empirically illustrates how smart-city data can potentially influence the power dynamics between actors, with unpredictable consequences. While planners and commercial actors perceived smart knowledge to be useful in optimising the travel patterns of cars, dissatisfied citizens argued that smart data should be assembled in such a way so that they could hold the providers of public transport accountable by measuring their level of punctuality. Citizens thus saw smart-city knowledge as a political tool that would enable them to influence public transport providers or their regulators. It is evident from this case that new knowledge devices can become subject to controversies due to the repercussions that new ways of knowing have for the power dynamics among actors within an urban system, and for the rationalities that inform systems governance.

Opportunities for re-coding urban knowledge production are not only related to traditional interest-based politics, but also to the 'politics of expertise', as knowledge production is often controlled by planning and governance communities centred around a particular type of analytical expertise (see, for example, Chapter 2). The influence of professional expertise on urban knowledge production may constitute a barrier to alternative ways of knowing due to the benefits that experts receive from their privileged position. This is because, i.e. first the authority of professional expertise often depends on them framing their knowledge as politically neutral. Attempts to politicise the assumptions involved in expert knowledge are therefore likely to face resistance from such actors as it may weaken their epistemological authority. Second, professionals may have an interest in maintaining established definitions of system boundaries and system functions as their expertise may only apply, or appear most relevant, within particular framings of these boundaries and functions. The promotion of socio-economic price signals as a salient knowledge device in Danish heat governance (Chapter 6, this volume) provides an apt illustration of the politics of expertise. It shows how the construction of socio-economic price signals were predicated on a network of contentious analytical assumptions that were policed by a powerful expert community anchored in the national administration.

Other cases, however, illustrate that expert communities are not always successful in policing the boundaries and functions. This is exemplified in Chapter 4, which illustrates how the knowledge hegemony of a highway- planning community within the National Government administration in Denmark was undermined, as highway planning transgressed the city boundaries of Copenhagen. In this case, roads that were to be transformed into urban highways were occupied by social protest movements. These protesters succeeded in making the negative impact of highways on urban life visible as a relevant governance concern. They thereby weakened the analytical-system boundaries constructed by the highway-planning community, which had hitherto excluded the negative effect of highways on their surroundings from consideration. In other cases, expert-based knowledge even plays a creative and innovative role in reframing urban systems. In Copenhagen, for example, planners drew upon the nationally dominant socio-economic political rationality to frame urban cycling as a health-promoting mode of mobility. The importance of this reframing is that, under this socio-economic political rationality, it made the creation of cycling infrastructure a highly attractive governance action.

The cases in this book thus portray the production of urban knowledge as a political phenomenon that is being configured, controlled and transformed by complex assemblages that often interconnect different forms of legitimation, expertise and authority. This suggests that the transformation of political rationalities is predicated on the ability to analyse, challenge and rewire the relations that configure these politico-epistemic assemblages. A further central finding from the contributions to this book is that expertise constitutes an epistemic resource that is often instrumental in black-boxing knowledge politics, as it may be used to purify knowledge as neutral and analytical. Like all other ways of knowing, expert-driven knowledge production depends on assumptions for making complex and uncertain realities intelligible as structured, predictable and actionable.

How a knowledge-politics approach can support the governance of sustainability transitions

Most governance techniques proposed and discussed in the transition literature focus on defining and promoting alternatives to established system designs.

These types of governance techniques include, for example, niche experiments (Schot and Geels 2008), urban living labs (Voytenko *et al.* 2016) and comprehensive visions of transitions (Roorda *et al.* 2014, p. 27). A key function of these techniques is to provoke, politicise and destabilise otherwise path-dependent governance practices. Less attention has so far been given to conceptualise how such interventions may penetrate and transform normalised daily governance practices related to urban systems. We suggest that attention to the reciprocity between knowledge production and political rationalities offers a fertile and strategically relevant conceptualisation of some of the mechanisms that institutionalise everyday governance activities pertaining to urban systems. The knowledge-politics perspective opens up avenues for understanding transitions governance not only in the form of innovation by experimentation and jointly developing grand guiding visions, but also in the form of rather subtle, mundane activities of knowledge production. Embedding alternative ways of knowing in devices and procedures for everyday knowledge production thus offers a transition governance approach that does not operate by openly confronting an established system design with a new one. In contrast, it operates by systematically accustoming actors to knowing urban systems in new ways and hence reconfiguring the knowledge that they have at their disposal. The cases presented in this book illustrate that devices and procedures for producing, using and (de)legitimising particular ways of knowing urban systems are never completely stabilised, but also never completely fluid. Knowledge production always plays out on a somewhat intermediate level of institutionalisation or durability and – if you wish – structuredness. This makes experimenting with new ways of knowing urban systems so relevant to urban governance, particularly in relation to complex new challenges like sustainability.

Finally, we have seen in the chapters of this book that close-up analyses of knowledge production as a productive political ingredient in urban transition governance can benefit much from governmentality studies (Miller and Rose 1990) and science and technology studies (Latour 1999), both in a conceptual and in a methodological sense. The governmentality literature, and its productive conceptualisation of power, provides an analytical lens for appreciating the inseparability of the formation of governable subjects and objects and the knowledge-producing techniques of modern governance. In addition to this, science and technology studies help to conceptualise how knowledge-producing techniques are typically wired into wider politico-epistemic assemblages, while at the same time appreciating the precariousness and mutability of such assemblages. Ways of delineating urban systems, as well as ways of producing knowledge about them, are thus usually contested – for the very reason that they define, constrain and shape environments for agency and interventions. Yet in shaping these environments, agency is often distributed among diverse actors. In most cases, non-governmental actors participate in these knowledge politics, or at least try to. To acknowledge how precarious the stability of seemingly entrenched knowledge assemblages can be and that, in principle, networks of non-incumbent actors may often find possibilities to challenge them, is of course not in the interest of those benefitting from an established way of knowing. Incumbent actors hence often take strategic action to obscure these possibilities. Knowledge politics will hence always involve experimentation, and it principally yields potential for altering power dynamics as a consequence of creative epistemological and discursive action.

Notes

1 Department of Planning, Aalborg University, Denmark.
2 Department of Urban and Rural Development, Swedish University of Agricultural Sciences, Sweden.
3 Institute of Environmental Social Sciences and Geography, University of Freiburg, Germany.

References

Chambers, R., 1997. *Whose reality counts? Putting the first last*. London: ITDG Publishing.

Flyvbjerg, B., 1998. *Rationality and power: Democracy in practice*. London: University of Chicago Press.

Jasanoff, S., 2005. *Designs on nature: Science and cemocracy in Europe and the United States*. Princeton, NJ: Princeton University Press.

Laswell, H., 1935. *Politics: Who gets what, when, how*. London: McGraw-Hill.

Latour, B., 1999. Circulating reference: sampling the soil in the Amazon forest. In B. Latour, ed. *Pandora's hope: Essays on the reality of science studies*. Harvard, MA: Harvard University Press, pp. 24–79.

Miller, P. and Rose, N., 1990. Governing economic life. *Economy and Society*, 19 (1), 1–31.

Partzsch, L., 2017. Biofuel research: perceptions of power and transition. *Energy, Sustainability and Society*, 7 (1), 1–10.

Patterson, J., Schulz, K., Vervoort, J., van der Hel, S., Widerberg, O., Adler, C., Hurlbert, M., Anderton, K., Sethi, M. and Barau, A., 2017. Exploring the governance and politics of transformations towards sustainability. *Environmental Innovation and Societal Transitions*, 24, 1–16.

Roorda, C., Wittmayer, J., Henneman, P., van Steenbergen, F., Frantzeskaki, N. and Loorbach, D., 2014. Transition management in the urban context. *Guidance Manual*, 49.

Schot, J. and Geels, F.W., 2008. Strategic niche management and sustainable innovation journeys: theory, findings, research agenda, and policy. *Technology Analysis & Strategic Management*, 20 (5), 537–554.

Temper, L., Walter, M., Rodriguez, I., Kothari, A., and Turhan, E., 2018. A perspective on radical transformations to sustainability: resistances, movements and alternatives. *Sustainability Science*, 13 (3), 1–18.

Voytenko, Y., McCormick, K., Evans, J. and Schliwa, G., 2016. Urban living labs for sustainability and low carbon cities in Europe: towards a research agenda. *Journal of Cleaner Production*, 123, 45–54.

Index

Page numbers in **bold** denote tables, those in *italics* denote figures.

For Product Safety Concerns and Information please contact our
EU representative GPSR@taylorandfrancis.com Taylor & Francis
Verlag GmbH, Kaufingerstraße 24, 80331 München, Germany